One Life

RICHARD E. LEAKEY

An Autobiography

To Meave, Anna, Louise and Samira

One Life

RICHARD E. LEAKEY

An Autobiography

MICHAEL JOSEPH
LONDON

First published in Great Britain by
Michael Joseph Limited
44 Bedford Square, London WC1
1983

ISBN 0 7181 2247 X

This book was designed and produced by
George Rainbird Limited
40 Park Street, London W1Y 4DE

Text set by SX Composing Ltd
Rayleigh, Essex, England
Illustrations originated by
Adroit Photo Litho Ltd
Birmingham, England
Printed and bound by
Mackays of Chatham, Kent, England

CONTENTS

LIST OF PLATES

PRELUDE

Most of us have had at least one incident in our lives that is permanently imprinted on our memories. In my case it was an unusual one that lasted only a moment. It was a happy, even joyous moment although I doubt whether I showed any outward sign of the emotion I was feeling. I was looking at a glass (or it might have been plastic) bottle. It contained blood-stained fluid, my blood mingled with urine – my own urine. Strange as it may seem, this was the cause of my joy. It was the first time in about three months that I had been able to pass urine and now, once again, this normally simple action was within my capabilities. I was in London; it was 29 November, late in the evening; I had just come around from an anaesthetic, following a lengthy operation during which I had received a kidney transplant from my younger brother Philip. Although I knew that there could be problems ahead, at least I was alive, and the transplanted organ had begun to function. I felt that one life was over and I now faced a completely new one – I had just been reborn. With that thought, I slipped back into the comfort of a heavily drugged sleep and some hours later I re-awoke to begin my new life. That, however, is not what this book is about.

My operation had been performed at St Thomas' Hospital in London. I had arrived in England with Meave, my wife, on 14 July 1979, for what I hoped would be a short visit. Indeed, I was so sure that I would be back in Kenya by the end of the month that I travelled with nothing more than an overnight bag. We had left our children with Meave's sister and her family, who at the time were stay in our home in Nairobi. The fact that Meave insisted on coming to London should have alerted me to the seriousness of my condition but I was far too confident, or too sick, to even think that I might be in danger of dying.

My doctor in Nairobi had advised me to travel to England because of recurring trouble from high blood pressure which had reached a stage that was impossible to control without special drugs – at the time unavailable in Kenya . The blood-pressure problem, caused by a malfunction of both my kidneys, had already resulted in two stays in hospital during the previous eight weeks. In my mind, however, I was convinced that I could keep my kidneys functioning a little longer simply by taking drugs to reduce the high blood pressure. I was desperate to avoid giving way to my illness; there was so much I still wanted to achieve. After our arrival in London, I went to see the specialist to whom I had been referred and I was told the worst: my visit to London was to be indefinite. A return to Kenya would depend upon either a successful kidney transplant or else a satisfactory adaptation to a home dialysis regime. It was made clear that I had reached the end of my particular road and that I could no longer live a normal life. Indeed, it began to dawn on me that I might have no life, normal or restricted, left to live.

Although, in the depths of my mind, I had known that such a day of reckoning was inevitable, I had always managed to put this knowledge aside, but to hear the medical verdict so clearly stated in Meave's presence, was a terrible blow. Whether it was the realization and acceptance of my condition or whether it was the natural course of my illness I am not sure, but my health certainly deteriorated very rapidly thereafter. Within a week I began to have serious difficulties: walking and climbing stairs was impossible; the days merged with the nights; the high urea content of my blood clouded my mind.

The first dialysis was surely one of the most extraordinary experiences I have ever been through. It was necessary for the blood lines from the kidney machine to be attached to blood vessels deep in my groin. Locating these was extremely painful. The dialysis lasted for several hours and I felt progressively better as each hour passed and for the first time I began to realize just how unwell I had been. As time went on, I began to feel increasingly cheerful, renewed in spirit as much as in blood. At last I was released and told I could go home until the next dialysis in two days' time. To my utter dismay, I was simply unable to spring off the bed and dress myself. I was a physical wreck. With great effort and much help from the nurses and Meave, I got myself into a wheelchair and my wife took me to a waiting cab. It was the first time that I had ever been in a wheelchair and been grateful for it.

The euphoria of feeling so much better from the first dialysis wore off quite soon and after the second day I was ready for the reinsertion of tubes in my groin. The second occasion was much worse than the first because the blood vessels were now bruised and sore, and the third was excruciating. However, by the end of the latter, six days after the first, the doctors were able to use a vein in my arm. What a relief it was! I felt comfortable; there was no pain and with fairly careful organization I found it was perfectly possible to do something else at the same time. With the help and tolerance of the nursing staff I arranged for my thrice weekly dialysis to be in a small side room off the main renal ward and this gave me the chance to sit at a table and write. I managed a regular and fairly voluminous correspondence but this was not enough to keep me from being bored, so I decided to write about my life.

CHAPTER ONE

AN AFRICAN CHILDHOOD

MY LIFE BEGAN SOMETIME IN 1944 at the Nairobi Hospital in Kenya, where I was born in the early hours of the morning of 19 December. As far as I know the event was normal although my mother has said that, on arrival, my bladder was full and the first involuntary action was to urinate on the doctor's face. Were my kidneys already trying to tell me something? Much has happened to me since that moment and I want to write about some of the things that I have done and enjoyed before they become blurred with time. I may, on occasion, have allowed fantasy to colour my story. Because I never kept a diary I can only write about what I remember, except where historical perspective has enabled me to understand a side of my country which, as a child, concerned me little.

The Kenya of 1944 was very different from the modern nation which has emerged since its independence in 1963. In the 1940s most African countries were still colonies of Britain, France, Belgium or Portugal – Kenya had fallen under the influence of the British Empire as a result of the country's potential for trade and settlement. Before the first British traders arrived on the scene there had been some trading by Arabs, who had settled along the East African coast as early as the late 9th century and who extracted ivory and slaves from the African hinterland. British trading, which began in the late 19th century under the auspices of the Imperial British East Africa Company, resulted in a series of treaties with the Arab rulers and the tribal chiefs inland. In 1893 the Company went bankrupt and the British Government was persuaded to take over the administration of British interests, which led in 1895 to the establishment of the East African Protectorate. Between 1896 and 1901 a railway was built from Mombasa to Kisumu linking the coast with the great lakes.

By the end of World War 1, European settlement was beginning to gain momentum and was actively encouraged. Alienation of land that belonged to Africans soon became the order of the day and the indigenous people were becoming increasingly fettered. The political domination of the majority by the European minority soon led to frictions, and by the early 1920s African opposition to colonial domination was firmly established. The major issue was that of land and agricultural policy, but no attempt was made to rectify the problems. The African population was widely regarded as a resource to be exploited for the good of the British Empire and any attempt to bring emancipation to the masses was held to be contrary to the Europeans' best interest.

Political agitation was quiet during World War 2 but considerable resentment was generated in the post-war period when the Crown was giving out rewards. Although countless Africans, including Kenyans, had died while serving Britain in campaigns both in Africa and abroad, African soldiers were given a cash reward of paltry value. In contrast European servicemen were entitled to easy terms on land grants, government loans and the like. All this gave considerable impetus to the nationalist movement, called the Kenya African Union, which became an active political party championing the cause of self rule. By the early fifties a policy of violence was introduced and terrorism became a common occurrence. The organized campaign to dislodge British rule was known as the Mau Mau, and in 1953 a number of important nationalist leaders were jailed – one of those incarcerated was Jomo Kenyatta, who later became Kenya's first President. Kenya at that time was unpleasant; a State of Emergency was in force between 1951 and 1960 and during this period my father was given a heavy police guard. Although only a child, I was very conscious of the fact that some terrible incident might affect our lives.

By 1958, it was clear to all that Independence could not be denied and a new attitude began to develop in the country. Jomo Kenyatta was released in 1961, and before long constitutional talks were under way to bring about Kenya's political independence which was finally granted on 12 December 1963.

The Kenya into which I was born was still, therefore, a British colony with a large and domineering settler community of European farmers, many of whom were former British servicemen or the sons and daughters of English landed gentry. Most saw the country as theirs by right and they looked upon the indigenous African population as a source of cheap though 'inefficient' labour. Some efforts were made to improve the Africans' standard of living, but it was generally felt that 'the African' was happier left to live a simple life. Few colonials made any attempt to learn or understand the way of life of the local people or to learn their language, so it is hardly surprising that servants were regarded as inefficient.

Kenya is a land of great diversity and contrast; although the equator bisects the country the climate is extremely varied depending on the altitude. The extremes, of course, are the hot and humid coastal belt at sea level and the frozen glaciers and snowfields on Mount Kenya at 17,058 feet

(5194 metres). Although the country covers just under 250,000 square miles – an area about the size of Texas or a little less than three times the size of England – at least 65 per cent of the land is arid semi-desert and unsuitable for farming. However, much of the high land is excellent for agriculture and the early European settlers were quick to appreciate this. The Kikuyu, who live in the central highlands, are one of Kenya's largest ethnic groups and by tradition farmers. They were particularly affected by the policy of land alienation which required that the indigenous populace be restricted to Native Reserves; a good part of the British settlement was on land formerly owned by the Kikuyu. The Masai managed to retain a large part of the country although not always the same area that they had traditionally occupied. They were persuaded to leave vast tracts of fertile land and to accept a larger acreage in the less fertile southern Rift Valley; the prosperity of these people was quick to decline.

The farms of the European settlers were usually large, often thousands of acres, and the basic labour force was provided by the local people who were seen as squatters. They were allowed to live on and cultivate a small piece of land owned by the European farmer in return for a greatly reduced cash wage. It was generally believed that the African did not need money provided he was able to fill his belly with the most basic food! In contrast, the Native Reserves in which the majority of the indigenous Kenyan people lived were ridiculously small. In these areas, large-scale farming was not practised and land ownership was vested in the tribe rather than in an individual which meant that land could not be used as financial security and as a means of obtaining loans. In much of Kenya, the boundaries of the Native Reserves were so restricted that very serious overcrowding and land abuse resulted. The rest of Kenya was considered either Masai land, game sanctuary, or simply uninhabited.

There are at least fifty tribes in Kenya, some more closely related than others but each having distinct traditions and cultural affinities. The colonial administration saw that it was to their advantage to maintain these differences and the major tribal groups were actively encouraged to distrust each other. Because the strength of unity as a strong nationalist tool was fully appreciated, a policy of 'divide and rule' was advocated.

Although the settlers too easily allowed an artificial barrier to come down between them and the people in whose country they were unwanted guests, there were exceptions. The Christian Churches ran schools in the so-called Native Reserves and prepared the people for 'civilized (in other words European) life'! There was a difference in attitude between the established churches and the missionaries: the missionary movement

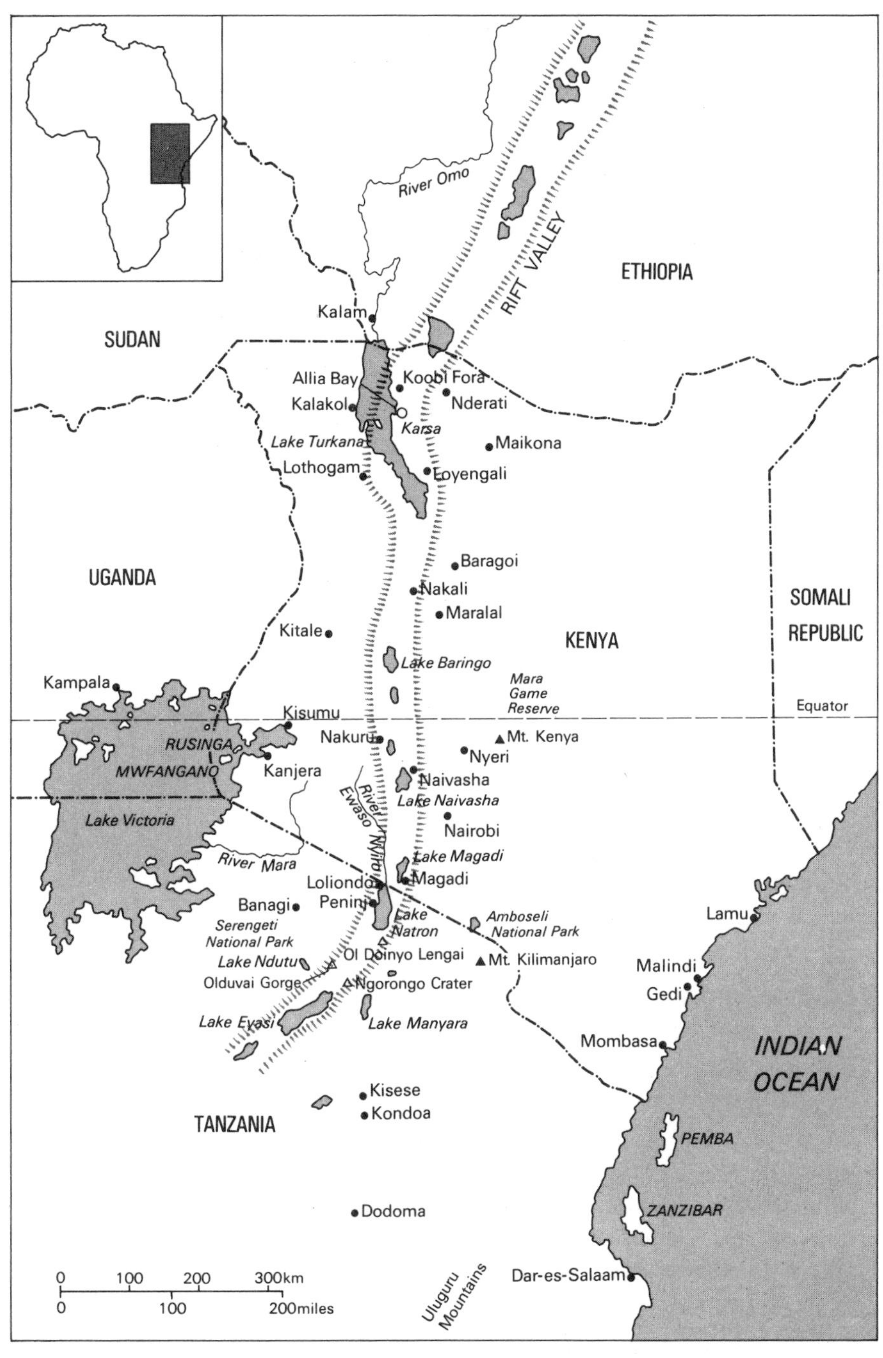

Part of East Africa showing the places mentioned in the text

fraternized and worked with Africans while the established churches were strictly for the Europeans. Racial segregation was official policy in Kenya then, as it sadly still is in South Africa, but in some ways Africans were worse off in 1944 than those in South Africa today because it was almost impossible to arouse an effective opposition. Before the days of television, it was easy for the European nations to be 'unaware' of the plight of the African in Africa.

In those days, the terms white and black were not used; one was either European, Asian, Arab or African. Business and trade were in the hands of Europeans and Asians. The Asian traders were disliked by both European and African and they, in turn, poorly treated Africans. The Kenyan Asians occupied a variety of other positions: some were in prominent posts, some were artisans, and the lower echelons of the Civil Service were mainly filled by Goan Asians. Goans were considered to be superior to other Asians since they at least had a Christian background, even though it was Catholic, and many had some European blood. There were also Asians who worked closely with African Nationalists, and the more prominent Asians got on well with their European counterparts. In general, however, there was a colour bar and the indigenous African was certainly a fourth-class citizen occupying the lowest rung of the ladder. But here again a good mission background was seen as preferable to the dreadful prospect of an African who was educated in an independent or private school. The latter was seen as frankly subversive – he had an idea of his own identity and even thought he had rights!

One of the positive consequences of the colonial administration was the promotion of Kiswahili, the language of the coastal Waswahili people, as a national *lingua franca*. The colonial administrators all had to learn Kiswahili, and it soon became widespread. The welding of a united Kenya after Independence from such diverse indigenous elements would have been considerably more difficult had there not been a common African language with which to communicate. During my childhood I could converse in Kiswahili with people of any tribe, whereas during my father's boyhood most of the people beyond the coast only spoke their mother tongue.

The towns of Kenya were very crude and, with the exception of Nairobi and Mombasa, they were rather unattractive. Many of the buildings were made of sheet iron on wood frames; the streets were earth and the general appearance was of dilapidation and dust. Nairobi, as the administrative capital of the colony and the trading centre for much of eastern Africa, was more imposing although it lacked any grand noteworthy buildings.

Mombasa was different from Nairobi. It was an active port with a long history of invasion and occupation by foreigners. It existed long before the Europeans arrived, and the people had their own traditional way of life. Many of its people were Islamic, spoke or at least read Arabic, and had an allegiance to the Sultan of Zanzibar who was respected by the British Government. The colonial authorities, therefore, treated the people of the coast differently; there was more social contact between the races although it was still rather superficial.

Such was the society in which my family lived. My paternal grandparents were Anglican missionaries and my parents were scientists in a country of colonials who really had little time for either. My grandmother, Mary Bazett, had first been to Kenya in 1888 as a young girl. As a good Victorian, from a family with strong church connections, she wanted only to serve her God and her country. After falling sick in Mombasa she returned to England where she married Harry Leakey, then a young curate, and returned with him to Kenya. On her second visit, she settled with my grandfather near Nairobi at a place called Kabete where a mission of the Church Missionary Society had been established a few years before. I never knew my grandparents; my grandfather died in 1940 before I was born and my grandmother died in 1948 before I was old enough to appreciate who and what she was.

My father, Louis, was born at Kabete in 1903. He grew up there, helping with his parents' work at the mission and being taught to read and write by governesses. From early boyhood, it seems, he developed the interests that led him into a career in anthropology and archaeology. Apart from a couple of years at a small school in England during 1911–12 when his father was ill, it was not until he was seventeen that Louis attended a proper school. He was then sent off to Britain to obtain a solid English education.

Partly because of this childhood, my father grew up with social attitudes that were definitely not typical of Kenya's European community. As a youth, many if not most of his friends were Africans. There were very few European children to play with near the mission, or indeed in Kenya. He grew up with Kikuyu boys and was even accepted as an initiate of the tribe. My father acquired a complete mastery of the Kikuyu language and he was also fluent in Kiswahili. Few Europeans have ever been so completely familiar with the Kikuyu customs as he and, although it is not well known, some of his most valuable writing was published as a three-volume study of the Kikuyu. My father was immensely proud to be a 'white

African', as he regarded himself. As a result of his early training and interest, he became a scientist rather than taking up a career as a missionary. This background made him different from conventionally educated Europeans and he did not fit easily in English society.

Before I was sent to school, I was fortunate in spending a lot of time out-of-doors. Both my parents believed strongly in taking children along on their field trips – if it was possible. Among my earliest recollections, are long bumpy journeys over very rough and dusty roads. In those days, there were no roadside bars or places from which refreshments could be bought and I particularly remember the introduction into Kenya of fizzy drinks. The first of these, purple in colour and known as Vimto, precipitated considerable argument in our family. But despite my protests, my father always preferred to stop by the roadside and boil water for coffee or tea on a small kindling fire, rather than buy new-fangled bottled drinks.

I have virtually no recollection of the first few years of my life but I am quite sure that my very early exposure to the African bush had a major influence on me. To this day, I love to spend time in the wild desolate places which unhappily I get to less and less. Nevertheless, even the briefest moments away from the sounds and smells of our technological world are a tonic to me; I feel cleansed and recharged.

I wish that I could remember more of my early childhood but, as for most people, it is the first scars and accidents that remain with me most vividly. One which I remember well occurred in the grounds of the Museum, where we lived. I was playing with some other children and one of the older boys, perhaps a friend of my brother Jonathan, who is four years my senior, gave me a ride on a child's tricycle. I was propelled at what seemed a tremendous speed until the tricycle was quite out of control and I landed in a very prickly bougainvillea bush. The sensation of being on a moving vehicle over which I have no control is now only too familiar and perhaps it was this first fright that gave me my life-long dislike of trains and other people's driving! I pass that same bougainvillea every morning when I drive into the Museum grounds.

In the beginning, the Coryndon Museum in Nairobi where my father worked and which was later to become the National Museum of Kenya was a small building that consisted of one main gallery with two floors. In the late forties and after the war, my father was involved in raising money through public appeal to add an extension to hold six new galleries. Although I cannot remember it clearly, I know that one of the things I enjoyed then was wandering through the building site after the workmen

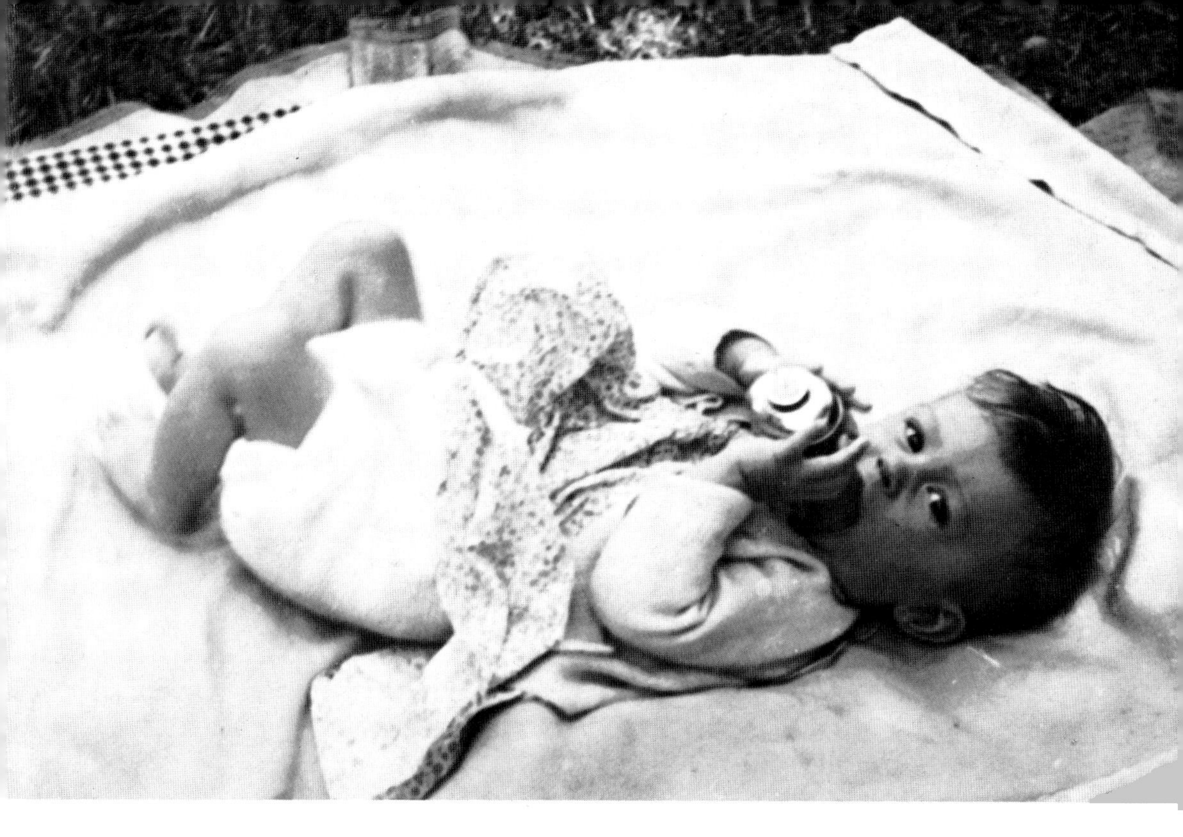

ABOVE Before I could walk

LEFT My mother with me and a dalmatian

BELOW In a pram, while Jonathan poses for the photographer

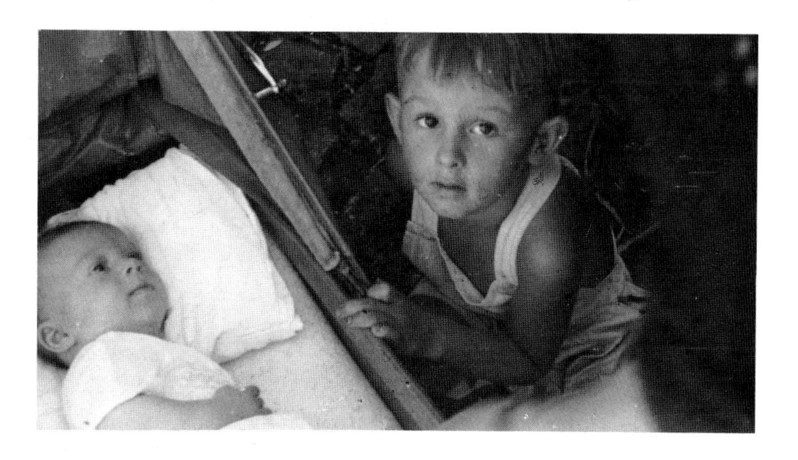

LEFT My first car was a serious matter

BELOW With Jonathan at play

BOTTOM On safari we were bathed in natural pools, which were often very cold

had gone home. It was rather like playing in a ruined castle as we engaged in games of hide-and-seek or waded through the sand piles. We were not supposed to go barefoot because of the jigger mites which lay sacs of eggs on the soles of the feet and toenails.

During the early years of my life, we lived in a small, stone house which was the official residence for the Museum's Curator, a post which my father took up in 1946. This house replaced a rat-infested corrugated-iron construction where we initially lived and which continued to function as a store after our new house was built. I clearly recall the rat hunts with the dogs on Sundays, when thirty or forty of these rodents might be killed.

One incident that I remember took place in the Museum house. My brother Jonathan and I were enthusiastically learning to make plaster of Paris casts of fish with a standard museum technique that had caught our imagination. The technique was quite simple and great fun. A fish, procured through parental goodwill, was laid to rest on a bed of sand in a suitable container – often one of my mother's baking dishes. The sand was built up to provide a perfect line so that exactly half the fish was buried, and half exposed. We poured molten wax over the fish giving it a coat at least a quarter of an inch thick. Once the wax had cooled and hardened, the dish was turned over, the sand washed away and the fish extracted from the wax mould. Freshly mixed plaster of Paris was then poured into the mould and allowed to harden for several hours. The whole thing was then placed in a container of boiling water to melt the wax, and the plaster fish was extracted. When the water cooled the wax was re-used for the next cast.

We used to boil our wax in the kitchen, and on this particular occasion I had a large saucepan on the wood stove containing several big chunks of brown wax being slowly melted. I went out to see my brother, who was in the garden, and for some reason forgot all about the wax on the stove. I am not sure how long it took for the wax to melt but it finally boiled, overflowing on to the wood stove where it immediately burst into flames. My first sight was of flames and smoke billowing from the kitchen window. I immediately ran to tell my father that the house had caught fire and then I waited for the inevitable. Fortunately, the fire kept to the kitchen and within a very short space of time it had been put out but, quite rightly, I was thoroughly thrashed. The house still stands and is now occupied by the offices of the Wildlife Clubs of Kenya and I drive past it whenever I park my car in the Museum's grounds.

All children are prone to be a nuisance to working parents and I am quite certain that I was no exception. One of the advantages that my parents had and, indeed, many parents have even to this day in Kenya, is that it is easy to employ someone to look after their children so that they can pursue their own work. These nursemaids were a characteristic of the former British colonies where labour was cheap. Many children throughout the length and breadth of Africa were brought up by women other than their own mothers. In my case, the lady who looked after me until I was about five was called Adijah and I remember her with great affection; I do not know from which tribe she came.

As a child, it seemed to me that my parents were away a great deal at certain times of the year although this was probably not so. At the time I could not understand why it was not always possible for my brothers and me to go with our parents on their early expeditions to Olduvai. Of course, I realize now that water was very scarce and my father could not have taken children who would drastically increase the consumption of the precious commodity while being of no benefit to the expedition. As a result we were left with several different European caretakers, who ran the house in my parents' absence.

Adijah used to spend hours with us while our parents were working. She taught us to speak and understand Kiswahili from a very early age; initially I was more articulate in Kiswahili than in English, simply as a consequence of having spent more of my waking time with someone who spoke no English. As well as Adijah I also learned from a number of other Kenyan employees who worked for the family in one capacity or another; to me they were a part of the family although some anomalies troubled me. For example, Adijah, so close and so comforting to me, had a very different relationship with my mother and father. She ate separately, never joined in social events, lived alone in a rather small and dingy room outside our home, and was very subservient. Was this because she was an African or was it because she was a servant? I do not know how far my powers of reason could have extended at the age of four, but I was certainly aware of these differences. In those days racial barriers existed as a social norm but I was confused, especially as my father was so obviously not tied by conventional attitudes in his relationship with his African friends.

In addition to Adijah and the domestic staff that my parents employed, there were many employees under my father by virtue of his job as Museum Curator. All of these people were kind. I had many hours of pleasure being spoilt by them, so that a strong bond of affection and trust developed. As I grew older, I think I accepted the idea that the social barriers existed more because I was the boss's son rather than because of the lack of pigment in my skin.

My father often used to tell us stories of his own unique childhood in Kenya. He used to regale us with folk tales, accounts of his hunting exploits, his training as a young warrior and so much more. It was always obvious that father was an African who just happened to be of a lighter colour. This, of course, added to my confusion. Looking back today, I feel that my father was an African, but one who was perhaps a little formal and westernized. Throughout his life he maintained very close relationships with a number of his childhood friends, yet he was also very inconsiderate and insensitive to many of his African employees. This puzzled me but, on reflection, it probably had more to do with a typically British, middle-class attitude towards the 'working class' than anything else.

Life became more serious in my fifth year when a baby brother, Philip, appeared and when I began to attend school. My first year of schooling was at the local Loreto Convent in Nairobi. In those days in Kenya there were no kindergartens or nursery schools but the Convent seemed willing to have a sort of playgroup. Although I was not a Catholic child, the nuns were very kind to me, and I enjoyed myself immensely. My earliest recollection of school was watching Jesus swing on a rosary and wondering how I could get such a glorious toy.

Perhaps this was the period when I first became aware of the fact that there were two sexes. There were many little girls in the school and I established a clear preference for the company of members of the opposite sex which I retain to this day. My stay with the nuns was brief because I was soon moved to St Mary's, another Catholic school but specifically for boys, and after only a short period I went on to the Nairobi Primary School. Although my mother had been a Catholic, neither of my parents practised any form of religion and I cannot imagine why they sent us to church schools except, perhaps, that they may have wanted us as adults to have the freedom to accept or reject Christianity. Certainly our family life had no element of religion in it.

I was at St Mary's the day that the death of England's King George VI was announced. We were all herded into the chapel to pray. The deep grief of the older boys and staff made a strong impression on me and heightened my awareness of the existence of a British Empire. I remember being puzzled that the local African people were not really responding to the King's death in the same way and, even more surprisingly, my European elders did not seem to expect them to.

There was a great deal of praying to be done over the death of the King and school life was interrupted for several days. Fortunately for me, most of the services were for the Catholic pupils and those of us who were not Catholic could stay away. This gave a number of us the chance to make our own direct contribution to the public mourning. In my case, this involved making a fine model of a church from cardboard, stones, and other salvaged material. In my church I put a number of praying mantids with a view to their doing their part for the Empire in its hour of grief. Unfortunately, the insects did not live up to their name and instead of praying, the larger mantids ate the smaller ones and I was accused and punished for being disrespectful to God, Church, King and country!

The days at the Nairobi Primary School were the happiest school days. I spent seven years there, doing well in the 'A' stream and always coming out in the first three places in the end-of-term examinations. Parental pride in my success was obvious and gave me tremendous pleasure.

It was at the Nairobi Primary School that I did my one and only piece of stage acting. The school play was an enactment of the Coronation of Queen Elizabeth II which had taken place in Britain the year before, and I had, for reasons I never knew, been selected to play a lead role as the Archbishop of Canterbury. Perhaps it was because two of my uncles were real bishops; anyway, rehearsals went on for several weeks and I began to enjoy the whole thing. There was, however, one action that appalled me – the Archbishop had to kiss the ring on the hand of the Queen! This I found terribly embarrassing and I always managed to avoid actually doing the fearful act at rehearsals. On the day of the play, when all parents and other dignitaries were to be present, the prospect of kissing the girl's hand proved too much, and I became ill, leaving the role to my less prudish stand-in. It was many years before I fully recovered from this social impediment!

CHAPTER TWO

'GO AND FIND YOUR OWN BONE'

As a family the Leakeys have always relished challenges. Even as a small child the trips I was taken on were real adventures. They usually involved travelling to areas where my parents had archaeological work. During those post-war years funds were very scarce and the expeditions tended to be short in duration and primitively equipped. We were accompanied by a few Kenyan employees and the family's dogs. My father had only his small salary from the Museum and a few grants from foundations and trusts in England, such as the Charles Boise Fund, that never amounted to more than a few hundred pounds a year. Nothing was wasted, food had to be eaten to the last scrap. Although we were extremely poor compared with the average European family in Kenya we were nevertheless much better off than the majority of Kenyans, including those who worked for us.

In the late 1940s my father became particularly interested in the various Miocene sites in western Kenya around the shores and on some of the islands of Lake Victoria. Miocene is the name given by geologists to a period in the earth's history between about 26 and 7 million years ago. The deposits in western Kenya were laid down between 20 and 15 million years ago, and were of interest because they were of the right age in which to find the fossilized remains of early apes and other primates that could throw light on the very beginnings of the long story of human evolution.

The visits to Lake Victoria, each lasting about a month, were particularly exciting. For several weeks before setting off, we all felt we were preparing for an adventure. Lists were made, remade and discussed. Supplies were assembled and packed into rugged wooden boxes. Plans were made, and on the day of departure we would be awake long before dawn. Most of the loading would have been completed the day before but there were always a few last-minute adjustments including arranging for the bedding to be laid out on top of the other loads in the back of the old Dodge box-bodied car. Mother and father sat in front and we children (my brothers Jonathan and Philip and various friends) were bundled in atop the rest of the luggage, often comfortable but with no room to sit upright.

Father always liked an early start and usually by sunrise we were well on our way. The Dodge had a wooden body with a wire cage, and it was because of this open-sided arrangement that we were able to feel so much part of the passing landscape. In those days there were few cars on the road and it was exhilarating to drive slowly through the countryside, breathing the early morning fragrances: the cedar-wood smoke from the homesteads, the smell of cattle, sheep and goats as we passed by villages, and the dank but wonderful smell of a forest, dripping with mist in the light of a Kenya dawn.

The first major stop on the journey was a small farming town known as Naivasha, some 100 miles north-west of Nairobi where petrol was available. This part of the road was very good, having been constructed by Italian prisoners of war held in Kenya during World War 2. Past Naivasha the road continued along the bottom of the Rift Valley for about another hundred miles to another farming town, Nakuru, where we would again stop for fuel. As a rule we would be in Nakuru in time to buy fresh bread, sausage rolls and other delights for a picnic lunch before starting on the rough section of the trip where the roads became little better than tracks strewn with boulders and pitted with ruts.

One picnic that stays in my memory was enlivened by a hopeless chase after one of my brother Jonathan's pets – a species of long-tailed field mouse that he had had for some time. It was presumably delighted to escape in the long grass of the picnic spot. However, the family rallied round and a search ensued but, of course, our little friend had disappeared. This was one of the first of what was to be a great number of childhood tragedies involving wild animal pets.

Our initial destination, Kisumu on the north-eastern shore of Lake Victoria, was reached sometime during the night. My father had a large wooden cabin cruiser or motor launch for work on the Miocene fossil sites and it was kept moored near the Shell fuel jetty at the Kisumu docks. The port is an important terminal on the lake and before the current political identities of Kenya, Uganda and Tanzania were so strongly developed, there was a great deal of water-borne traffic to and from it as a convenient point of access to the hinterland and, via the railway, to the seaport of Mombasa.

The car would be unloaded during the night, and in a relatively short space of time everything would be carefully stored on the old boat, known appropriately as the *Miocene Lady*. She was 44 feet overall and the design was typical of the larger motor launches of pre-war vintage. There was a main cabin with the captain's seat and wheel set before the port bunk. The crew's quarters were forward of the cabin, while aft there was a small galley with two Primus stoves. There were twin engines, large Ford V8s with wooden covers, which also served as rather warm third and fourth bunks. The after deck was open except for a tarpaulin cover supported by a single spine, the tarpaulin itself being braced by slats tied down along each side. My parents occupied the main cabin while we three children were accommodated with an extra bed set aft of the engine bunks.

Those night-time arrivals at Kisumu were always intensely exciting. The loading and stowing had to be done by the boat lights and this, of course, meant that the engines had to be kept running. The movement of the boat, the noise of engines, and my father's commands all created an atmosphere of urgency and expectation in which we children revelled and it was not without some sadness that finally the word was given and we were on our way. Father liked to travel the seven- or eight-hour journey to the fossil sites in the night or early morning to avoid the stronger winds and heavier seas that are a characteristic of the hot and often humid late mornings and afternoons. Father employed a Kenyan captain, Hassan Kabiru, who knew the lake like the back of his own hand. This was reassuring when overcast skies made the lake dark and forbidding.

I particularly enjoyed being awake on the boat in the early hours before dawn. As we made our way along the lake, a cool wind blew and, periodically, a falling star punctuated the inky sky. By sunrise, usually glorious with warm red skies and golden clouds, we were well on our way and father would ask Hassan to take the *Miocene Lady* alongside one of the many fishing canoes which would be taking in their gill nets. We would then start the day with marvellous fresh fish for breakfast. To taste pan-fried fish in the perfection of an African dawn must be one of the most delightful of all experiences. Sometimes, less happily for me, father would cook his special kedgeree. This consisted of rice, chopped hard-boiled eggs and pieces of cooked fish – and I loathed it.

The fishing on Lake Victoria continues today with modern canoes driven by small engines, but in those days when I first knew the lake, many of the canoes were hand-made: planks of wood stitched or tied together with bark rope. They were often as much as 40 feet long and were propelled by paddles: five or six muscular men could move these craft at an amazing speed. Some of the canoes had a short mast for a characteristic lateen sail and with the right wind they were very fast. The fishermen used simple gill nets which were left out overnight, anchored and buoyed. To avoid over-fishing the government quite rightly imposed a control on the size of mesh that could be legally used. Inevitably, fishermen would cheat in their desire to improve their catch and quite often, as Hassan changed course to go alongside a canoe, we would be horrified to see the nets and fish being thrown overboard and the canoe make off. The fishermen feared that we were a patrol vessel and they were presumably dumping any incriminating evidence. Such a waste, and we all felt badly for differing reasons. There were also large craft, rather similar to the typical East African coastal jahazi or small dhow, which were used to transport freight and passengers between the major towns around this huge inland sea.

Our destination on these visits to Lake Victoria varied but from 1947 through to the early fifties my parents had a lot to do on the two larger islands, Rusinga and Mwfangano, at the mouth of the Winam Gulf. Both islands have extensive deposits of Miocene sediment and each season many fossils were collected following the regular heavy rains which in this part of Africa come twice a year.

The fossil bones are found there because the remains of animals that lived during Miocene times became buried as a result of the regular and continuing process of sediment accumulation in these African lakes. Each year for millions of years, layer upon layer of silt, sand or mud has built up, trapping animal remains. Given time, new erosion has re-exposed the layers, and whenever there is rain or heavy wind new fossils are uncovered. This makes fossil hunting a long-term business because every year there could be something new.

My father's boat, the *Miocene Lady*, was the base for operations and it would be moored in a suitable protected bay. While the family lived on the boat, a camp was established on shore for extra people in the party such as scientific colleagues, who came along at their own expense, and the Kenyan excavators.

One of the first rituals that I really looked forward to was a bathe in the lake. We were taken ashore in a small dinghy and my father would carefully scan the bay with his binoculars before blasting off with both barrels of his shotgun into the water where we were to bathe. As soon as the echoes of the blast died away, we were allowed to rush into the water and have our swim while father stood guard. The noise was to scare the many crocodiles. My father believed that it would take about five minutes for these flesh-eating reptiles to recover their equanimity and, by this time, we would, it was hoped, be safely ashore again.

On one occasion, my father decided to shoot a particularly large crocodile that appeared singularly unmoved by the noise of his gun. It was well known to the local villagers and greatly feared. We children stayed on the boat and watched while father, with his rifle, carefully stalked the beast.

The hunt was successful and we were taken ashore to see the crocodile, which measured just under 16 feet from head to tail. Father wanted the belly skin so, assisted by some of his workers, he began to skin out the carcass. It was then that a most remarkable thing happened.

Gradually, we all became aware of a loud and continuous rustling in the grass and bush nearby and upon investigation, we discovered a vast army of 'siafu', the infamous safari ant of Africa, advancing menacingly towards the beach, presumably alerted by the smell of blood. The belly skin of the crocodile was removed just in time and as we left, the first ants had begun to swarm over the carcass. These ants are vicious so we had to get clear by wading into the shallows. The next day, we returned to find the crocodile, which had probably weighed close to 2000 lb, reduced to scaly skin and bones. All the softer tissues had been devoured! These same ants are known to kill young animals, including calves and lambs, and sometimes children.

In the mornings we generally accompanied our parents to the sites, and while they worked we would play nearby, usually under a shady tree. The many local children would gather to watch these strange people at work so that sometimes we would get to play with children of our own age. I remember being taught to make model canoes and boats from the reeds, sticking each reed to the next by driving long acacia thorns into it much as you would use nails to join planks. The creativity of the local children was infinite – I often think we handicap the children of technological society by giving them too many ready-made toys and detailed instructions with models and games.

At noon or, if we could persuade our parents, sooner, we would return to the boat. Often we had a long walk of several miles along the narrow footpaths that are so typical of these islands. As our procession moved through the fields, greetings would be exchanged with all whom we met and passed. It was a tremendously warm and friendly place and I especially enjoyed the creased smiles of the old ladies working with hoes in their cotton fields. Often, an old lady would have a long-stemmed tobacco pipe clenched between her toothless jaws. Women in Africa carry extraordinary loads balanced on their heads and as a result their posture is magnificent.

The rest period varied but as children we seldom slept and preferred to fish from the boat, using a home-made rod and line baited with flour dough. On most days, we would catch dozens of a species of *Tilapia* that was too small and bony to fillet but was fine in a stew much liked by the Kenyan workers. I was happy to be able to make such a gift of fish to my friends ashore who seemed to have a much less comfortable existence than we did. This fishing was exciting and I used to compete with my brother Jonathan. Once I was so excited by my success that I fell overboard and had to be rescued by Hassan who saw the incident from ashore. It was the first time I had been in water out of my depth and even then I seemed to have had an instinct for survival – I kept myself afloat by treading water. Hassan reached me long before Jonathan had been able to rouse our sleeping parents. To some, it might seem that I have always been accident prone but I do not think that is the case. It is just that I have lived a fairly active life in some quite wild places and misadventures do occur.

As a general rule, father used the cool afternoons and evenings to search for new sites and to re-examine old localities. These walks were great fun not only because of the lower temperature but also because he had the time to tell us about the natural history as we went. Throughout his life, father studied East African wildlife and he even found time to publish two books on the animals of the region. The islands in the lake had considerable forest and bush cover with a tremendous variety of birds, and we became ardent naturalists with special interest in the butterflies and beetles. These poor creatures were gleefully caught in nets and by other means and put to death in our cyanide bottles.

The possession of guns was quite strictly controlled but Jonathan had an air rifle and was keen to make a comprehensive collection of bird skins and eggs. This gave us occasion to go off hunting and I am ashamed to say a good number of birds were shot. We were all able to prepare study-skins and many happy hours were devoted to hunting for the scarcer species which when killed were skinned and stuffed. Some of the birds proved very good eating and we would frequently kill a few extra pigeons to provide enough for a barbecue. Much later, when I left school, this interest in birds and insects proved useful because I was able to earn some money as a part-time collector for museums abroad. I now find it quite extraordinary that I could actually have enjoyed killing mammals and birds, and today I spend much time preventing small boys from doing exactly what I used to do!

In 1948, during one of those trips to Lake Victoria, there was a particular discovery that created a great deal of excitement. My mother found a remarkable complete fossil skull and jaw of a primitive ape; the first such skull to be found. No other skulls of this creature have been found since and it ranks as one of the most important finds ever made by my mother in her long and successful career. Although I was only four years old I still have a vivid picture of the site, which was on a small stony hill, quite close to a gnarled and virtually leafless tree. I remember the tree because the excavations took a number of days to complete, and during this time I had no option but to amuse myself as my parents worked. The only available shade was some way off from the digging where I felt very isolated. How I wished that the wretched little tree could have had leaves so that I might have a place to play near the spot where my parents were clearly having such a good time!

The fossil skull belonged to an ape-like animal which lived in tropical Africa some 20 million years ago. Prior to mother's find it had been known from fossil fragments of jaws, teeth, and a few limb bones but the skull, which is so important to scientists, was completely unknown. This primitive ape, known as *Proconsul africanus*, is still considered to be an important part of the long story of human evolution. In 1948 when mother found the skull, there was apparently a general feeling that she had discovered the 'missing link' between the great apes and man. One of the things that I have explained in my book, *Origins*, is that there is, even now, some uncertainty about the relationship of *Proconsul* to later African apes. Although a great deal more material is now available, including the best part of a skeleton, there is some argument as to the evolutionary place of this animal in the broad picture of our own past. All the same, the *Proconsul* skull is of critical importance and another like it has yet to be found anywhere. There is a great deal that is still unknown about primate, and particularly ape, evolution between the time of *Proconsul* some 17 million years ago and the first upright apes, about 4 million years ago.

My father, as might be imagined, was immensely excited. This find alone was ample reward for the very difficult times my parents had endured without recognition or encouragement. It was the first really important fossil to have been recovered in Kenya and it provided the opportunity for my father to raise money for further excavations.

In 1948 Nairobi was certainly not a centre for studies on human evolution and so it was considered essential to send the skull to England where the great scholars in the subject could examine it after the formal descriptive paper had been written. My mother, as the finder, was given the honour and task of taking it. My father had a long-term view of life and a strong sense of Kenya's ultimate destiny; after discussion with the authorities he agreed to send the skull to England on loan, on condition that the specimen remained the property of Kenya and could be recalled at will. In 1970, I requested the British authorities to return the skull but by that time the records had been lost and somebody at the British Museum had actually registered the fossil as British Museum property. It was only in 1981, thirty-two years after my mother took the specimen to England, that we finally located the original 1948 correspondence and the skull was returned to Kenya.

I found my first fossil bone, I think, in 1950 at the age of six. I had been told very firmly by my parents not to get in the way or distract them from a delicate excavation they were conducting. We were at a site called Kanjera on the shores of Lake Victoria. The concentration shown by my father on this particular occasion suggested that we would be late in breaking off for lunch. I was miserable. There were flies everywhere and there was no shade. What could I do? I was forbidden to help them and unable to go off alone because of the dangers of snakes and other wild animals in the surrounding bush. My response was typical of a six-year-old. I set up a persistent whine, hoping to create sufficient irritation to bring about a change in the *status quo*. 'I am hot', 'I am thirsty', 'I am bored', 'I have a stomach ache', 'What can I do?', 'What time is it?', and so I went on.

The parental response was quick and unexpected. I was told by my father to find my own bone and dig it up! I was partly pleased, but would have preferred a swim in the cold waters of the lake. There are always fragments of bones lying on the surface of the fossil sites, most of which are unimportant and can be dug up without any harm to the scientific record. So, I moved off to look for my bone – at least I had something to do! I had only gone about 30 feet from my parents when I found a small scrap of brown-coloured fossil bone showing on the surface of the site. I had my own dental picks and small brush and I set to work, probably with some reluctance.

It proved riveting. My bone quickly showed signs of being large, well embedded in the sediment and, more important, the shiny enamel surfaces of some teeth quickly appeared after my initial scraping and brushing. What was it? I become engrossed and was soon entirely oblivious of the heat and flies for the first time that day. This is now a common sensation but then it was my first experience of the incomparable thrill of uncovering something that has lain buried for hundreds of thousands of years.

My prolonged silence soon aroused my parents' curiosity. When a child, especially a little boy of my disposition, is consistently quiet for more than a few minutes, there must be an excellent reason. My father came over and was instantly alert. I was happily uncovering a complete jaw of an extinct species of giant pig, the first such complete specimen to have been found! I was quickly sent off to find myself a less important bone while my parents abandoned their original find and set to work on mine. I was furious and deeply upset. None of their compliments about my discovery had the slightest effect in relieving my feelings!

Every time I see this specimen in the collection of Kenya's National Museum, I remember the incident. Indeed, I often wonder if it contributed to my original firm decision to avoid at all costs a profession that involved excavation and the search for fossils!

In the early fifties, partly because funds were short and perhaps because of the growing security problems in Kenya as a result of Mau Mau activities, my father turned his attention to sites in Tanzania (then Tanganyika) at Olduvai and also to the south of Olduvai where there are many important prehistoric rock paintings. He first went to Olduvai in the twenties and it became a favourite site especially after the discovery of some hand axes in 1931. My mother is still working there. The safaris to Tanganyika were quite different from those to the hot humid sites in western Kenya and they exposed us to life in the dry country. In 1951 my parents made a three-month trip to central Tanganyika to study the rock art in the Kondoa area and, once the school term was ended, my two brothers and I were able to join them. The result of my parents' work in 1951 was a large collection of accurate colour reproductions of the prehistoric art that abounds in this region of East Africa. These reproductions are in the collection of the Nairobi Museum and some accurate replicas of the paintings form a permanent display in the prehistory gallery.

Prehistoric art is widespread in Africa. It is known from southern Africa, Zimbabwe, Tanzania, Ethiopia and North Africa. The Tanzanian rock art has much in common with that in Zimbabwe and in southern Africa but the typical polychrome paintings found in these areas are not found in Tanzania. These rock paintings are enigmatic in so far as it has not been possible to date them accurately or to identify the artists with certainty. Sadly the paintings themselves cannot be directly dated. But pigments similar to those used in the artwork were recovered in some excavations at the sites from which carbon dates were obtained.

During my parents' project several excavations at various rock shelters were conducted and a lot of archaeological material was recovered. At one site at Kisese, 20 feet of deposit were excavated and the date of deposits from the upper nine feet indicated that the lower levels had accumulated over 29,000 years ago, perhaps as much as 50,000 years ago. Pieces of red ochre were found in almost every layer, and colouring pencils as well as ochre-stained palettes and pebbles used for crushing were found intermittently from levels older than 29,000 years. It is likely that the artists were painting several thousand and perhaps many thousand years ago, although the rock art that we see today cannot be that old. Indeed, the tradition of rock art is known to have continued in parts of Africa until quite recent times.

Some of the Tanzanian rock paintings are extraordinarily beautiful. The artists managed to depict the essential points of an animal with a few simple lines producing remarkably life-like portrayals. They frequently made use of the natural contours of the rock face, resulting in an almost three-dimensional image of an animal. People were commonly painted and some of the activity scenes are unrivalled: elephant hunts, ceremonies and dances. Two scenes in particular are of special interest to me and both are shown in the Museum exhibit. One of these depicts a woman being pulled by her arms in opposite directions by what are obviously two competing pairs of suitors! What was the event that fixed itself so vividly in the artist's mind? The other scene that I enjoy is a group of bathers who are frolicking or dancing in a river. The strong implication of real pleasure and happiness is a delight to see and shows that these prehistoric people had a great capacity for joy and were not miserable communities struggling for survival.

Why did prehistoric people paint on the walls of caves and rock shelters? Were they simply passing idle hours or did the paintings have a spiritual or magical meaning? Perhaps they were a method of passing messages and news between groups of people who were always on the move but who rarely met. At the moment no one can say for sure. We can only stand in wonder before an art developed many thousands of years ago by Africans in Africa.

In 1951 when my parents began to study these cave paintings the journey from Nairobi took several exciting days with many stops to cool overheated radiators, mend punctures and dig ourselves out of sand or mud. It was always amusing to be told that we were following the Great North Road with which the British planned to link Cape Town and Cairo. The dream of a major north-south road remains elusive to this day.

Upon arrival in the Kondoa District we made our way to a Government Rest House at Chungai which was close to a number of rock shelters. My father would always call in to see the local administrator before establishing us in our new home. The Rest House, as I remember it, was a typical bungalow with an outside lavatory and kitchen. The house was surrounded by a wide verandah which, along with the windows, was screened to keep out the flies and mosquitoes. The compound consisted of various outbuildings. Nobody told me how the local people managed without mosquito protection; there was no evidence of screens on their houses.

Though these sort of questions were not large in my young mind I was already beginning to worry about my privileged position as a European.

The Kondoa area was not particularly rich in wildlife when we were there but it obviously had been. Records made by European travellers indicate that there was, as late as 1930, a great variety of large and wonderful animals. Elephant, rhino, buffalo, eland, kudu and countless others roamed the wide valleys and *Brachystegia* woodland. All these large mammals were shot and all the vegetation cut down by the colonial administration in an effort to eradicate the tsetse fly from the area and so eliminate sleeping sickness. Can anything be more tragic? Literally thousands of animals and much of the vegetation were destroyed without anyone protesting and the tsetse fly continued to thrive as before, feeding off small mammals, reptiles and birds. The decimation of Africa's wildlife, a direct result of the ignorance and folly of the Europeans, must stand as a tragic legacy of the colonial era in Africa.

In the 1950s, the roads in the Kondoa area were blocked at intervals by tsetse fly control points. These were large wooden barns built across the road into which every car had to be driven. The doors were closed behind the car and a couple of men would then spray everyone and everything with insecticides, probably DDT, for several minutes. This exercise was designed to prevent flies being transported by vehicles from one place to another. I cannot believe that it was very effective but for children it was tremendously exciting and made the world of adults seem so important and responsible. It is not so amusing when one is grown-up and I remember being extremely indignant a number of years later when at Johannesburg the South Africans felt it necessary to disinfect the aircraft on which I was travelling plus its human occupants. Was it we Kenyans who were the object of their aerosol or were there really pests on board that might contaminate South Africa?

Although, as I have said, large game in the Kondoa area 30 years ago was already on the decline, there were still many animals that would come close to our bungalow much to our excitement and delight. For instance, about once a week my father used to slaughter a fat-tailed sheep to provide everyone with fresh meat. After the sheep had been killed and skinned, it was hung by its hind legs to age for a day or so under one of the many large trees in the Rest House compound. This always attracted a leopard and we would hear its rasping call as it prowled about outside the house. I remember having to be reassured that leopards in African never ate living humans! Because the bungalow was small and not especially cool at night, we all used to sleep on the verandah where I was aware that a mosquito gauze screen would not offer us much protection against the huge cats.

The other nocturnal visitor that created great excitement was a bush baby; a wonderfully cuddly primate with huge eyes and ears that is common over most of Africa. I used to put out a saucer of bread and condensed milk or some fruit and try desperately to remain awake so that I could see this little animal come to eat. What pleasure I had when I succeeded!

Snakes were common in the area, so my parents made a rule that we boys should always wear long khaki-drill trousers. These were hot, and for boys who had always worn shorts, they were most oppressive. My father's reasoning was that the most likely place for a snake to bite would be our legs and thick material would deflect the fangs, thus saving us from certain death. Fortunately his theory never was tested. In any case the popular myth that snakes are always looking for a chance to bite a human is far from the truth. Almost always if a snake bites it is in self-defence. What would *you* do if someone trod on you while you were peacefully sleeping in the sun? Most snakes have a natural fear of man, and make off as soon as they are disturbed.

Notwithstanding, an unexpected encounter with a snake can make one forget such wise remarks! I remember that once my father was leading us in single file along a narrow path through towering elephant grass when he put his foot on what he thought was a log. The log turned out to be a python, at least 18 feet long, stretched out in the sun, trying to digest a small antelope it had just swallowed whole. The python, quite naturally objected to having its postprandial snooze disturbed and let out a tremendous hiss. The noise was so obviously reptilian and hostile that it sent us all scampering back down the footpath while poor father had to creep back past the swollen, angry reptile. However, neither my brothers nor I have any fear of snakes and, indeed, today Jonathan actually runs a snake farm.

My father was entitled to 'home leave' from the Museum every four years. This meant that, periodically, the whole family was taken off to England. This 'leave' was a relic of the Victorian belief that it was essential for English people to get away from the tropical sun now and again to avoid creeping madness or worse still, becoming completely 'African'. To 'go native' or to admit to *not* longing for the joys of English civilization was a terrible stigma that ruined many a promising career.

I remember little of those visits to England except being amazed by the extent of the damage to London following the German bombing raids and

ABOVE On Lake Victoria the *Miocene Lady* was often our home

LEFT Father's camps were always simple

BELOW One of father's early expeditions to Tanzania – these were always an adventure

ABOVE Getting to know my younger brother

LEFT Our nurse, Adijah, with Jonathan and me

BELOW Long shorts were very comfortable – if not stylish

I was immensely thankful that we were not there then. As I grow older I become more and more opposed to the nuclear arms build-up but equally, I value freedom too much to believe other than that we must always be prepared to fight to preserve it.

On my first visit to the Tower of London when aged five, I was thrilled by the sight of the Beefeaters in their red uniforms but the concept of locking up princes and generally being beastly to other Europeans or, worse still, fellow English, was a shock. My colonial childhood had given me the idea that all Europeans stuck together and never had to be jailed or executed.

My parents also took me to see what seemed to be an endless number of cathedrals and, as a result, I developed a passionate and lasting dislike for ecclesiastical architecture. The smallness of England and its neat hedge-bound fields were unimpressive to a small boy from Africa. For me the best part of these visits to England were the journeys there and back. In 1947 I must have been one of the last children to have flown from Kenya to England in one of the 'C' Class Empire flying boats operated by British Overseas Airways. I only had one such journey because about two months after our return they were replaced by land planes. The flying boat began its slow journey at Lake Naivasha and arrived at Southampton three days later, having stopped on the way at Kisumu, Entebbe, Juba, Khartoum, Alexandria, and several European cities.

On one of these early childhood 'home leave' trips, we spent some weeks in France in the Dordogne and the Pyrenees. These two areas, apart from being the most beautiful in France, are of course where so many of the famous European cave paintings have been discovered. My mother had spent a good part of her own childhood in France with her father who was an artist and she has always loved the Dordogne and the wilder Pyrenean valleys of southern France. I shall always recall my first experience of the Dordogne caves, deep underground, mysterious and hauntingly beautiful. In the strange light shed by our acetylene miners' lamps, we followed narrow passages and tunnels into caverns of stalagmites and stalactites. Here and there lifelike animals and mysterious signs and symbols were painted or engraved on the walls. Most impressive of all was Lascaux which had only recently been discovered and was at the time being studied by the Abbé Breuil. Caves, with their tunnels and crevices, stalactites and stalagmites, always seem to be fascinating and mysterious places for children. And here there were the dramatic colours of the paintings themselves. This visit to France must remain one of the most extraordinary experiences of my childhood.

CHAPTER THREE

CAUGHT IN MY OWN TRAP

IN 1955 I COMPLETED my primary education and went to a secondary school. The move from one school to the other coincided with the family moving from their house at the Museum to my parents' home that they had designed and constructed over the preceding three years. Like many people, my father wanted a house of his own which would also be better suited for dogs and children. But his salary from the Museum was rather pitiful so he had to wait until he could raise the money to buy some property and put up a house himself. The new house was built around a courtyard and was located on five acres of land some twelve miles from Nairobi in an area known as Langata, which was then just a collection of separate residences and dirt roads. My mother still lives there when she is in Nairobi.

My new school, known as the Duke of York, was a fairly new boys' school conveniently situated on the western or Langata side of Nairobi. Although I was not a boarder, I belonged to a House and I was expected to take part fully in the routine of school life. The regime at the Duke of York in 1956 was typical of British public schools in the Victorian tradition, with considerable emphasis being placed on strict discipline. Though I am glad this system of education no longer exists in Kenya, I have to say that I think I benefited from the strict discipline.

My parents held their children on a very loose rein, and they were far too busy with their important work to fawn on their children as most parents do. I suppose they must have talked to me about fossils but I was more interested in the natural history of living things. It was very much a working household. My father would work late at the Museum and then write and study at home. My mother was busy too, and I always seemed to be at school, doing homework or generally messing around. Therefore, my brothers and I developed very independent characters. In my own case I had to learn that independence, though essential to me, needed to be tempered with self-discipline and a preparedness to accept responsibility for my actions. These attitudes I learned at the Duke of York School.

My first day there began rather badly. I was dropped off early by my father and as a day scholar I did not see my brother Jonathan who was then a boarder. Like any boy at a new school, I was very shy and tried desperately not to be at all obvious as the older boys gathered for assembly. Unfortunately, one of the boys who had been with me at the Nairobi Primary School and who did not particularly like me, decided to gain favour from the older boys by pointing me out as an object of potential interest. I was put forward as a known 'lover of niggers' (I had, I suppose, at some time identified my abhorrence of the racial attitudes of those days, when Kenya was still strictly segregated). Anyway, the little beast of a boy was an immediate success with his senior fellows who set upon me with glee. I confess it was the first time I felt real terror.

Before I knew what had happened, I had been placed inside a wire cage some three cubic feet in size. The cage was of a kind that was then used for transporting large glass bottles of chemicals well packed in straw. The hinged lid was closed and padlocked. I was crouched like a monkey in this tiny cage, with no way of escape. The boys were delighted and several hundred of them took great pleasure in teasing me through the wire. I was poked with sticks, spat upon and even urinated upon, for what seemed an eternity until it was time for everyone to go to assembly. I remained in my cage, very miserable and frightened. After assembly, the boys went off to class and I was eventually released by a senior master who had to get a hacksaw to cut through the padlock. He had no doubt that I was to blame; so, wet through, filthy and stinking, I began my first day of senior school.

A few days later I was again in trouble. It was after lunch in the House and we juniors were supposed to be quietly doing some reading before beginning afternoon lessons. Several older boys called me over and I was told to sit in a chair so that they might question me. Word was obviously out that I was easily bullied for, as I sat down, the chair was pulled away and I naturally sat on the floor. A bowl of tea had been placed in exactly the right place and I got my rear end soaked with sticky brown liquid. The boys roared with laughter but quickly dispersed when a House prefect appeared. As I stood there with dripping wet shorts it was quite clear to the prefect that it was all my fault and he gave me six strokes of the cane for ragging during prep time! I began to get the idea of how the world is run!

Clearly this sort of environment was not at all conducive to study and I began to have problems with lessons. I especially disliked Latin and maths, and I never mastered either. My class position dropped and my parents were dismayed. I never told them about bullying at school; this would have been frowned on as 'sneaking' and not keeping 'a stiff upper lip'.

Gradually I began to learn how to keep out of trouble, but inevitably as there were so many rules and regulations I was always finding I had unwittingly done something wrong. For instance, stocking tops had to be exactly three fingers below the knee, and mine were usually wrinkled around my ankles!

All the boys at school were European or 'white' and many were the sons of colonial settlers. Their attitude towards the indigenous Africans was totally contrary to mine and troubled me deeply. During this period Kenya was still suffering from the so-called Mau Mau, and much publicity was given in the newspapers and on the radio to the violence of the struggle. Indeed, the daughter and wife of my father's uncle, Gray Leakey, were murdered and he was taken hostage and died. We were not a close family, however, and I only came to hear of him at the time of his death. My father also had a price on his head for a time because he had been asked for help in resolving the problem by the British Government, and some extremists believed he was a danger to their cause. My unwillingness, therefore, to join in the blanket condemnation of Africans won me a special status and I was frequently bullied and taunted as a consequence.

My school career was also not advanced by my dislike of team sports. I was an incompetent games player, but some days there was a shortage of willing bodies and I was forced to play. I played cricket once, but only for a few moments because the first ball to be bowled at me hit my forehead and I was knocked out cold! Hockey was also a disaster: a fast ball hit my face, smashing my right cheek bone and rendering me unfit to play for weeks. The one game in which I once did rather well was rugby but it was, I regret, purely by accident. I had been press-ganged to play in the House 'B' team and while I was standing stupidly on the pitch as everyone rushed about, someone foolishly threw me the ball. Quite by chance I caught it but then I immediately realized my situation; I was now the target of the opposing team and I was sure to get tackled and, probably, hurt. My natural animal instinct was to run to save myself and this I did, fortunately in the right direction. As I sped down the pitch I was aware that I was being propelled entirely by fear of my pursuers. Eventually I tripped over and fell, still holding the ball. As it turned out this was just perfect, I had unintentionally scored what was to be the only try of the match! I was very careful to avoid rugby thereafter.

Although I frequently pretended to be ill to avoid games, I was in reality often far from well; my brothers and I were afflicted by a tropical disease known as bilharzia. This is contracted by wading or swimming in water containing snails infected with the parasite *Schistosomiasis mansoni* and it can be quite dangerous. I had three separate infections during my childhood and I commonly suffered from diarrhoea and headaches. My headaches may also have derived from a serious head injury which I suffered at the age of eleven, when I was thrown from my pony.

During this period of my childhood I began to think a little about religion. My mother is a once-upon-a-time Catholic and, as I have said, my father came from a deeply religious missionary and church of England background. (My uncle, indeed, became the Archbishop of East Africa in 1960.) My parents had decided that none of their children was to be christened as an infant though, of course, we had been sent to Church schools. They felt it was up to us to make or not to make the commitment to Christianity and I have always been grateful to them for this.

As might be expected, the school chaplain at Duke of York was an enthusiastic crusader who looked upon my lack of baptism as a personal affront. However, all his preaching was to no avail and I was often punished. Many 'free' afternoons were spent in 'detention', copying out parts of the Bible or Psalms. The whole exercise was so very petty that my position hardened. Even so I might have become a believer had it not been for one final injustice.

One day, I was on my way to school as usual when I was involved in an accident. Nothing serious, I simply fell off my motor scooter. My arms and legs were bruised and cut but, more serious, my motor scooter was damaged and would not go. As a consequence I missed morning chapel, arriving just as the boys were coming out, so I was noticed and reported. In spite of my excuses, amply confirmed by cuts and bruises, I was duly punished. The injustice of this incident led me to meditate more deeply about Christianity and to challenge many of the methods used in its teaching. I have never considered myself a Christian since.

My study of man's origins has, naturally, often led me to reflect about religion. I feel certain that religious beliefs are related to the dawning of self-awareness in primitive man. God is an inevitable creation of the human intellect where explanations are unavailable. We will never know when the God concept evolved but it was surely there 70,000 years ago when people began to practise burial rituals in the Middle East and Europe. Self-awareness and the ability to reflect on the universe may well be tied to the expansion of the brain seen in the fossil human ancestor, *Homo erectus*. Though we can never prove it, if this were so, 'religion' could be a million years old.

There are people who find a conflict between the scientific explanation of the world and their religious concepts. I see no reason why this should be so. Many leading scientists are deeply religious and it was, of course, the scholars from the great religious schools who made the earliest scientific experiments. I am not, however, a believer in the superiority of one religion over another. I am sure that religious elitism, a belief that some are 'saved' and others 'damned', is the basis of much evil, including racial prejudice. Jomo Kenyatta once noted that Christian missionaries had to break most of their ten commandments in order to establish themselves in our country.

I myself do not believe in a god who has or had a human form and to whom I owe my existence. I believe it is man who created God in his image and not the other way round; also I see no reason to believe in a life after death.

Having said all this, I cannot deny that I am nagged still by that big question 'What is Life?'. What is the 'thing' that is there in an organism one minute and gone the next? There is, I know, more to life than I understand and to me this is epitomized in that mysterious frontier between life and death – the 'spirit' as it were, that cannot be quantified but which is in living things both animal and plant and not in the dead. Perhaps science will one day give us the answer to this question too.

Much as I disliked school I had a full range of interests and activities that I was able to follow after classes, during the holidays and at weekends. The fact that I had so much to do out of school contributed to my poor performance in school. Indeed, I often played truant in order to get on with my own projects at home.

On one such occasion the form master had got flu and we were told to busy ourselves for the day on revision of past lessons. I sought permission to work in the school library but as soon as I was clear of the class room, I collected my things and made my way through the school compound to the adjacent forest. I slipped under the fence and walked about half a mile to the main road where I intended to hitchhike home. It was still quite early in the day and I planned to go off to the game reserve on my pony. As a general rule, school teachers are in school at the same time as their pupils, so I had no fear of being spotted on the road by a member of staff. Foolish optimist that I was! The first car to come along passed me and then slowed to a halt. I ran up, delighted with my good fortune only to find myself getting into the car of my form master, Mr Kitchener, who was a relative of Kitchener of Khartoum. He had gone to see the doctor and was on his way to collect some medicine from a chemist. Anyway, he took me home and the matter was not discussed on the journey. I knew that I would be losing face were I to seek clemency from Kitchener. He was a grand man and, like most of the masters who were former English secondary school teachers, in some cases from Eton, he hated boys who were weak in spirit. As it was, I had a rather good day: I did in fact do some revision although insufficient to help me in the test that I was given on return to school. The punishment I received on that occasion was certainly well deserved.

I had one other memorable encounter with Kitchener that was terrifying at the time though amusing in retrospect. A number of senior boys had made an underground study in one of the boarding houses. The excavations had been carefully conducted over a considerable period in order to avoid detection and the result was a small room measuring about 10 × 8 feet. Access was through a vertical shaft. The whole concept related to the craze that existed then for war stories, particularly those featuring the escape by British servicemen from German prisoner-of-war camps. The entrance to our underground room was concealed under a rug in one of the small studies that were available to senior boys and we used our hide-out for cigarette smoking and the consumption of illicit liquor, usually home-brewed beer or wine.

The usual procedure was that four boys would be down at any one time for a smoke while others kept watch in the study above. Generally the room was in use after lessons in the evening when I would be home but, because I was the main purchaser of cigarettes, I had certain 'rights' during the lunch and morning break. It was during a morning break that together with two others I was caught. We had been given away by another boy who was never identified although we had our suspicions. When this particular break-time was over we came up the shaft, emerging into the study to find Mr Kitchener waiting for us. For reasons that I do not fully understand to this day, punishment was less severe than I expected. We were not expelled but our subterranean room was filled in with stones and concrete along with all our cigarettes and wine. I would guess that Mr Kitchener felt that our enterprise was 'in the public school tradition' and the 'mole' or informer was to be despised.

I was attracted to horses very early in my childhood and learned to ride when I was aged seven. I thoroughly enjoyed riding and my parents agreed that I should have my own pony as soon as we moved from the Museum house to Langata. My parents were not particularly 'horsy' but both had ridden as children so they knew what was involved. Our problem was that ponies were not cheap and my father was far from being rich.

A long search led us eventually to a small animal that was broken but not schooled and which had been running free for several years on a farm near Nairobi. He was a marvellous animal, with a creamy-yellow body, a black mane and tail, and dark legs. When we got him we had neither stable nor paddock, and there was insufficient money to build both at once. This resulted in a compromise. The paddock was built first while the pony lived in the house in one of the rooms with an outside door! In those days lions were not uncommon in the suburbs and a pony outside at night would certainly have been eaten.

My joy at having a pony was intense and I took every chance to ride him. That he spent a good deal of time on two legs simply impressed me – once I had learned to hang on. The animal was very wild and I had to be very careful. I was soon confident enough to venture beyond the paddock and started taking longer rides in the neighbourhood along the rough earth roads. It was not long before I had my first experience of what became a familiar sensation: being on a pony with the bit between his teeth going full tilt and taking no notice of my attempts to slow him down. During one of these gallops I fell off, landed on my head and suffered a severe fracture. I was in bed for what seemed to be weeks and became thoroughly bored.

My mother took up riding again in an attempt to keep the pony exercised, and an Australian friend, Des Bartlett, who was chief photographer for the animal film maker Armand Denis, tried to school the animal for me. By the time I was allowed out of bed it had been decided that I had to have a more controllable animal. My second pony, Susie, was quite different and I began an extremely happy phase of my life – at last I had found something I was good at.

Susie was obedient, and with her I learned to show jump, to do dressage, and generally to participate in the full range of horsy activities. My parents were encouraged to help organize a new pony club at Langata and before long all the horsy children in the neighbourhood were members. The Pony Club was really more of a youth group: like all pony clubs, it was a wonderful opportunity for youngsters to do things together. We met every Sunday and so kept in contact with people whom we would otherwise seldom see because most houses were fairly distant from each other. The Langata Pony Club instituted all the rules and regulations that were passed on from the famous British Pony Club and in due course it was officially recognized by that body. We had regular instruction and frequent competitions and for the first time I began to enjoy competitive sport.

I bought another pony in my early teens and with this one, called

ABOVE First time on horseback; I was the youngest

LEFT Fish at Malindi. Hats were always worn in those days to avoid sunstroke

BELOW My older brother on a leash

FAR LEFT In an English garden

LEFT I hated being taken to museums; here we are seated on the steps of the British Museum, while on a visit to London

LEFT BELOW At the Tower of London, where the Beefeater appears to have impressed Jonathan

RIGHT With mother and Jonathan in the grounds of the Museum, Nairobi

BELOW Our camp in Tanganyika

ABOVE Probably the first family portrait, with some of my mother's dogs

LEFT In the bush at Kondoa, where my parents studied the rock paintings

BELOW My class at the Nairobi Primary School. I am at the back row, second from the left

Bonito, I competed in Junior Championship events. The pinnacle of my competition riding was the Junior Show Jumping Championship for East Africa, but to my shame I fell off in the third and final jump-off. This might have been because I had a high fever due to a bout of malaria at the time and so should not have been on horseback.

In addition to shows, I also enjoyed racing. This all began with a point-to-point when I rode a friend's racehorse over a steeplechase course. The horse was a fine animal and with little help from me it won. This success led to further opportunities and before long I was a regular rider in such events, although my growing teenage body soon became a major problem. Despite this I was able to ride professionally in flat races at the Nairobi Race Course and I had some thrilling wins. The owners and trainers were very kind to me but by the age of sixteen I had become just too big to compete. My last professional race was memorable. I was leading by several lengths, but as we came into the final corner before the home straight I just fell off! Why, I will never know, but it was an ignominious ending to a brief career as a jockey.

Another less orthodox sport that I greatly enjoyed was to chase and rope certain wild animals which abounded in the grassy plains just beyond my parents' home. This area is now dotted with small farms and homes but in the period before 1960 we could ride out over empty plains and within fifteen minutes of our stable we would find giraffes, wildebeest, and various antelopes. A little further afield were rhinos, elands, and, of course, all the carnivores that frequent the African savanna. Zebras were our favourite chase – they could not outrun a fit pony and were thus easy to gallop alongside and to rope. Wildebeest were also easy, but they had horns and it was necessary to work as a team. Giraffes were fun to chase but we never roped one. The animals were always released and it was all good sport although it could be dangerous.

Chasing rhinos was the ritual proof of teenage skill, and bravado. Rhinos, we discovered, will normally flee when frightened. Our aim was to get alongside a fleeing rhino, give the thundering beast a good slap on the rump which usually annoyed it so much that it turned and we would then have the excitement of being chased ourselves. We did not often achieve the slap and usually discretion dictated a hurried retreat!

These rides across the bush were tremendous fun and they also provided me with further opportunity to learn the ways of the wild. There were so many things to observe, to study and to enjoy that it is no small wonder that I became determined to have a career that gave me an outdoor life in Africa.

During my last few years at school, starting at around the age of thirteen, I made an income by capturing wild animals for our friend Des Bartlett. In the early 1960s, Armand and Michaela Denis were well known on British television for the 'Walt Disney' style of nature films which they made in Africa. They lived close to my parents' house and had quite an extensive set-up with studios in which they were able to do their own cutting and editing. Much of the filming was done using captive, semi-tame animals in controlled conditions so there were always lots of little animals about – it was like a small zoo and I became quite interested in it particularly after they had accepted a tortoise, and then a chameleon, which I had found. That was how I came to know Des and, through that friendship, I began to supply animals for him. Des would pay me the equivalent of about £10 for such animals as a mongoose, wild cat, porcupine, or bush baby. This sum may not have been much to the film industry but it seemed quite a lot to me and in this way I learned to handle quite a wide variety of animals.

I trapped the animals around the area in which we lived because, at the time, it was quite wild. The traps were set in the evening after school and any animals caught were collected first thing in the morning before I was taken to school soon after 7 a.m. Sometimes, when I left school early, I would take an animal down to Des, get paid and reset the trap. The money I received for my trapped animals was very important and gave me a marvellous degree of independence from my parents who could not possibly have afforded to give out large sums of cash.

There are many ways to catch wild animals, but for the small carnivores, the most effective one is to set out home-made box traps, baiting them with burned meat or dead birds. My main problem was that I had to be up extremely early to check my traps before I left for school.

A more exciting way of capturing animals was to borrow father's Land-Rover, which I had learned to drive illegally at the age of fourteen, and go out at night with torches and large nets, rather like those used for catching butterflies.

There were always things to be seen and it was a great thrill to be racing after a large mongoose at night as it scurried off down a forest track. The skill was to net the fleeing animal before it either took off into the bush or turned on you. Several school friends would join me and I probably laughed more on these evenings than at any other time in my life.

One particularly popular animal to chase was the African Spring Hare, which looks rather like a small kangaroo. They are nocturnal, live on the open plains, and feed away from their burrows. The method was to find an animal, using the Land-Rover lights, and then give chase on foot. Running across Africa in the dead of night after a hopping Spring Hare may not be everyone's idea of fun but it was for us and many of my Pony Club friends would opt for a Spring Hare hunt rather than spend an evening at the cinema or a dreary dance. As I recall, nobody actually broke a leg by falling into holes but there were a good number of tumbles and quite frequent encounters with larger animals including lion and cheetah. Apart from the one or two Spring Hares that I sold to Des, all were released immediately after capture.

My parents' house at Langata was always full of wild animals; additionally, Jonathan was already an avid collector of snakes, and had a small snake pit in the garden. I did not get on well with my elder brother and I probably went out of my way to irritate him. However, through him I often had the opportunity to handle reptiles and one of the things that used to excite me was learning to deal with poisonous snakes. Jonathan's interest was well known in the district in which we lived and many people who had problems with snakes would telephone or send a note asking him to help catch some offending serpent. The family association with the Museum also encouraged people to ask for assistance when strange animals were invading their homes.

More often than not, I was busy with my ponies, but if Jonathan was away the task of helping people usually fell to me. On one occasion – I think I must have been about sixteen – I received a distress call from a lady who reported that her favourite dog had been bitten by a very large puff adder; the dog was very ill and the snake was still there. Could I do something about it? I took my parents' small car – a Morris Traveller – and drove over to find that the dog was indeed very sick. I had brought our snakebite serum and gave the dog an injection in the hopes that it might save its life. I then went on to look for the snake which I found not far from where the dog had been bitten. It was partially concealed under some rotting timber. I was amazed: usually when people reported a large snake it turned out to be quite small, but this really was large! At least five feet long, it was one of the biggest examples of this species that I had ever seen, and certainly the largest that I have ever attempted to catch.

To catch a snake of this kind is not particularly difficult: one simply has to keep the dangerous end immobilized by grasping it behind the head. Once caught, it is easy enough to hold. The difficult task is to get it inside a bag where it can be kept safely. On this occasion I had brought a pillow case, but clearly it was not going to be big enough! I caught the snake without any difficulty and carried it to the car where I put it in the back intending to bag it later. I counted on the rather sluggish snake remaining still while I drove home: if it had moved I would have stopped and got out of the way. There were no problems on the drive and when I arrived I found a large bag and persuaded one of my father's staff to assist me. I recaught the snake from the back of the car and prepared to bag it. My assistant was holding the mouth of the bag open so that I could get the tail and body of the snake into the sack before letting go of its neck. But just as I let go, my assistant took fright and dropped the bag. On the spur of the moment, instead of letting the snake drop, I tried to grab it in mid-air. I brought my left hand across and caught the snake, but unfortunately I had it by the head. Before I knew it, it had bitten me firmly on two fingers.

I pushed the snake quickly into the bag and then I had to deal with my own snake bite. It is extremely difficult to give oneself a series of injections, but fortunately my father was in the house at the time. I rushed in to tell him of my predicament. He did not appear unduly worried but set to work with a razor, cutting close to the wound so as to cause bleeding. Whilst my hand was bleeding and the venom draining off, he went out to get the snakebite kit, which was still in the car. Fortunately, I had not given all the serum to the dog and my father gave me what was left. He injected it in my hand near the bite, as well as in my arm and chest. In less than an hour my arm had swollen to two or three times its normal size and my skin turned purple and black because of the bleeding in the tissues. The discoloration spread from my fingertips up to my neck and to the side of my face. The pain from the serum and the snake poison was by this time excruciating; I could feel every blood vessel in my arm as if each were red hot and burning through my flesh.

Neither I nor father felt that it was necessary for me to go to hospital immediately because all that the doctor could do we had done ourselves. I put myself to bed and hoped for the best. However, after the first twenty-four hours I began to worry that the wound might go gangrenous and so eventually I went along to the Nairobi Hospital where I asked for an injection of penicillin. In addition the doctor insisted on giving me a tetanus shot which happened to be a horse serum. This proved almost fatal.

Exactly ten days after I had been to the hospital, I went to bed early, feeling slightly unwell, only to awake at about ten o'clock with almost complete paralysis. My condition was such that I could neither get myself off my bed nor call out for somebody to come to help me. Fortunately, it passed after several hours and for the rest of the night I had no problem. The next evening I took the precaution of having some friends round to keep an eye on me as both my parents and brothers were away. Exactly the same thing happened, but the paralysis passed quite quickly and never occurred again. I did, however, develop very shaky hands and for a while these tremors were so bad that I was unable to pour out a cup of tea without spilling it. My shaky hands did not prevent me from learning to fly a plane although sometimes I used to find it very difficult to convince my passengers that I was not terrified by the whole business!

On a subsequent visit to England, I visited a hospital outside London where I underwent tests for a couple of days. It turned out that I had suffered some nerve damage caused by delayed shock due to the horse serum.

During my last year of school, when I was approaching my sixteenth birthday, the Langata suburb of Nairobi where we lived was occupied by a pride of seven young lions. The lions took pleasure in their new diet of chicken, ducks, sheep, cattle and horses. Every night or two some household would lose an animal to the lions. I was terribly excited and managed to get myself on the local gamewarden's team of rangers who went out every night on patrol. The idea was to keep up with the lions and keep them moving, thus forcing them to miss meals and so, we hoped, encourage them to go back to the nearby Nairobi National Park where wild animals were to be found in plenty. Night after night the lions managed to elude us, killing domestic animals with abandon while we searched elsewhere.

It was then decided to trap the lions so that they could be moved back to the park. My former experience at live trapping much smaller animals gave me some credibility, and before long I was, in effect, running the operation despite my youth. I had the use of two large traps, one with a wood frame and the other metal. Each afternoon, when I got back from school, the traps would be taken by the National Park's lorry to wherever the pride was known to be lying up. As dusk fell, I would bait the traps with large chunks of meat, often taken from the left-overs of the lions' kill of the previous day. I would then take up watch. The trap doors were to be released manually by me as soon as I had two lions in their boxes. To do this, I parked my vehicle about fifty yards away from the traps and attached two release pins to the vehicle by means of two long strings which were kept taut. I sat in the vehicle, watching the traps through binoculars ready to release the doors when my lions were in.

Frequently the lions would go off, not being in the slightest bit interested in our bait. Then the rangers and I would abandon the traps and try instead to stay with the pride through the night, so preventing them from killing. Often we failed and the next morning I would receive a call from an angry householder who had lost cattle or in one case 152 chickens!

Every night we saw hyenas, which would readily enter the traps and steal the bait. This was infuriating, but they helped to allay the lions' suspicions by entering the traps and leaving unharmed. The trouble was, by this time, the bait had gone and there was no reason for the lions to enter! I would rebait the traps but this usually disturbed things too much and another night passed in vain.

There were seven lions when the operation began, but two were shot – one was killed by the gamewarden in charge, and another by a nervous householder. In the former case, it was a question of shooting one lion to stop the others from attacking some horses when our usual shouts and displays were not enough to discourage the angry pride. The other lion was shot with a .22 rifle, a stupid and indeed dangerous action that could have had serious results. As it was, I received word from the game rangers that one of the lions was losing blood and had been abandoned by the pride. The local gamewarden was away for a day and I used this as an excuse to 'prove' myself. I decided to deal with the situation alone which, on reflection, was rather typical but nonetheless foolish.

I took my father's ancient 12-bore shotgun and, accompanied by a tracker, set out to follow the spoor of the wounded animal into dense bush. The trail was clear and we had no problems, crawling along, often on our knees, with good visibility forwards. I was undoubtedly preoccupied with showing how brave I was and I really had no clear idea just how precarious my position could become. Suddenly, my tracker and assistant grabbed at me from behind and pointed urgently ahead. There was the lion, a mere twenty feet away, looking for all the world as if it were about to spring. My reaction was, I regret to say, very slow, probably because my heart had stopped completely and I had no oxygen in my head! The lion remained motionless, as did we, and although I had it covered with a double-barrelled shotgun, I was reluctant to shoot. It seemed so quiet that it occurred to me that perhaps it was already dead. What a blow to my reputation if word got out that I had bravely shot a dead lion! After what seemed an eternity but was only a few moments, I whispered to my friend that he should throw a stone at the lion and, if it so much as twitched I would shoot. We threw dozens of rocks before being satisfied that indeed it was quite dead; it had been killed by the single .22 bullet which had torn the lungs causing fatal bleeding.

I managed to catch the remaining five lions over a period of about six weeks and I received quite a lot of publicity. The local newspapers were delighted to feature my exploits and on several occasions I was photographed looking proudly at a furious lion in its trap. The animals were released in the Nairobi Park and at Amboseli, the latter being a wildlife area some 188 miles away – far enough to ensure the lions would not return to the suburbs of Nairobi.

Apart from lions, I also once tried very hard to capture a wily leopard that frequented the Langata area and which was making a habit of eating dogs. Leopards are much more difficult to trap than lions, and I never did succeed. All the same I had a lot of fun and some quite amusing incidents.

I used to set the trap in the evening and check it in the early morning. After several weeks without success, I decided to modify the trap and bait it with a live goat. To do this, it was necessary to construct a leopard-proof section at the end of the trap; the leopard should enter being unaware that the wretched goat cannot be got at. The method is good at stopping hyenas stealing the bait because they are seldom attracted to live bait, but it was unfair on the goat. One night, I learned too well how the goat felt.

I had been out with a delightful teenage date for an evening in town and as I was but sixteen, I had had firm directions to have my girlfriend home before midnight and indeed to be home myself. My parents were away so my situation was less serious than hers. I had taken my father's little car despite the fact that I was too young to have a driving licence. No matter, the police in those days were far too busy with real crime and a European youth was seldom stopped on the road (another case of double standards).

As we neared home, I suddenly had what seemed a brilliant idea. I would visit my trap before taking the young lady home. I would leave her in the car while I went off, armed only with a torch, into the forest – what better way to prove my prowess! I told my girlfriend that if I were unduly delayed, she should take the car and go home and that I would walk the half mile to collect it in my own time. She was no fool and clearly thought that I was trying to impress her, as indeed I was. Accordingly, she drove off long before I could possibly have returned.

As I approached the trap, my clothes soaked by the dripping forest, I could hear the goat bleating its poor heart away and I suspected that a large carnivore was probably about. I reached the trap and shone my torch on the goat, looking for an explanation for its obvious unease. To my horror, I soon saw what it was: a leopard was crouched in the grass about twenty feet from where I stood and in the brief look I had it appeared to me to be quite angry. Again, fear froze my brain. I knew I had to escape and the obvious route was into the trap, closing the door behind me! This I did long before I realized that once trapped I would not get out until someone came to release me! There I remained, bleating with the goat until early in the morning when my absence from home alerted my father's staff and a search party set out. Needless to say, that was my last attempt to trap a leopard.

These out-of-school activities absorbed most of my energies and it is no wonder that I failed to do as well as I should have in my 'O'-level examinations. After four years at the Duke of York, I had no wish to spend a further two years preparing for a university, though equally I didn't want to annoy my parents by actually failing. In the event I narrowly missed a first-class pass which at the time was an essential criterion for further studies, and so could thoroughly enjoy my last days at school. It was now that I felt my real life would begin.

CHAPTER FOUR

ENCOUNTERS WITH LIONS

IN 1959 MY FRIENDS the Bartletts invited me to drive one of vehicles on a trip to the Serengeti. They planned to stop off at Olduvai, where my parents were working, because father wanted Des to photograph some of the work in progress. My father thought that some good photography and footage of cine film would not only publicize his work on early man but also enable him to lecture in America and to raise money. Afterwards I was to go along as an assistant to Des while he filmed some of the extraordinary wildlife in this magnificent area. This arrangement suited me very well because it was drawing close to the time when I would leave school, and I thought that I might even persuade Des to give me a job.

Olduvai Gorge, situated to the west of the eastern fork of the Rift Valley in Tanzania, has played an important role in the lives of my parents and their three children. The gorge itself, which was first recognized as a place of scientific interest in 1911, is 30 miles long and a place of great natural beauty. It cuts through a series of deposits, the earliest of which are close to two million years old, and in places is more than 300 feet deep, exposing old lake beds, stream channels and volcanic ashes. In the deposits are preserved fossils of many different kinds of animals as well as camp sites and the stone implements of early man. Occasionally, fragments of hominids have been found, too. Several of the volcanic beds can be dated directly by potassium-argon radiometric dating. In fact, Olduvai is the first archaeological site where this method was used, and the initial results suggesting that the earliest stone tools were 1.7 million years old were widely disbelieved. My father first began his work there in 1931 and my mother joined him in 1935. My parents have worked in the gorge ever since and, since the mid-1960s, my mother has virtually lived there.

In 1958 the trip to Olduvai was still a difficult one, and my parents were operating on an extremely restricted budget. Their camp was tiny and they slept in the large truck that my father drove and which, once at Olduvai, became the centre of the camp. A few tents were set up and the lorry was not driven until the return journey. Any essential trips, such as those for water, were made in the Land-Rover. Most of the fossil and archaeological sites were visited on foot which meant walking a considerable distance every time we went out; petrol was very scarce. In 1959 my parents had made a very important find; a single fossil human molar, clearly of great age. Father was quite certain that excavations would produce additional fossil material of this ancestor, and it was these excavations that Des was going to film.

For reasons that I do not fully recall, Des had had to postpone his departure by a few days and so we were late in getting to Olduvai. Rather than waste time, my parents decided to search other parts of the gorge and it was in fact the day before we arrived, 17 July 1958, that my mother made what must be one of the most significant discoveries ever. Although we were a healthy family, on this occasion father was in camp with a stomach complaint and my mother had gone off accompanied by her beloved Dalmatians. Towards noon, while working along the slope of the gorge at a site known as FLK (Frida Leakey's Korongo), she spotted pieces of bone washing out on the side of the gorge and these, on examination, proved to be parts of a hominid skull.

Mother hurried back to camp and broke the news to father. They both rushed to the site to confirm the reality of the discovery and then began to plan the recovery of the find. As it happened, we arrived the next day, finding their camp by looking for bits of paper tied to trees (a method they used as an aid for locating a camp); and the excavation of this vitally important specimen was all put on photographic record. Had we been on time, my mother might never have had the chance to search site FLK and so the skull of *Zinjanthropus*, as the find was named, might never have been found. It is ironic that my mother made her greatest discovery at a site named by my father many years before in honour of his first wife.

Up until that time my parents had worked the Olduvai Gorge with extraordinary conviction and with very meagre resources. Father was convinced that it was just a matter of time before they would find the makers of the many stone tools that are so abundant throughout the gorge. These stone tools occur in thousands, and are among the most important archaeological aspects of Olduvai and, although my mother has completed a great many excavations, many sites are untouched, to be excavated by future workers.

As it turned out, the 17 July discovery was not the maker of the many stone tools. Mother had noticed fragments from the base of a skull and a little careful brushing away of the earth revealed several enormous human-like teeth. It was in fact a complete set; subsequent excavations yielded the palate, the bones of the face and almost the entire skull which, although broken into pieces, was remarkably complete.

We arrived at Olduvai on the afternoon of 18 July, and plans were made to begin filming the very next day. Over the next ten days, careful excavations proceeded and Des took a very complete photographic record of every stage. As pieces of the skull were recovered they were taken back to camp where father would encourage my mother to devote her afternoons to sticking and reassembling them. It was soon clear that a truly remarkable find had been made.

When the skull was discovered, fossils of such age and completeness were very rare and it stimulated excitement among archaeologists all over the world. The skull belonged to a kind of early human known today as *Australopithecus boisei* – a side branch in the story of human evolution that became extinct about a million years ago. Shortly after the skull was discovered father published a paper describing it and he proposed that it be named *Zinjanthropus boisei*. (The Empire of Zinj was an ancient pre-colonial name for the East African coast. Thus the generic appellation means 'East African man', and *boisei* was in honour of my parents' friend, Charles Boise, who had encouraged and provided funds for their work in the 1950s.) In 1960 my father took the skull to the Pan African Congress held in West Africa. At that time it was often called 'Zinj' for short, and in my family it acquired the pet name 'Dear Boy'. In some press accounts and in popular books, because of its very large molars, it was called 'Nutcracker Man'. Its generic name, *Zinjanthropus*, was only used until it became apparent that there were so many similarities between it and material which had been found in South Africa that there was no justification in erecting a new genus for it.

The excavation that began under Des's camera continued for over a year and must be the largest ever conducted at Olduvai or any other early East African site. Des and his party, of which I remained a part, could not stay and we left on the tenth day to drive to Serengeti National Park where we were to meet Armand and Michaela Denis who were flying from Nairobi on the regular East African Airways Sunday day-trip. In fact we returned with Armand and Michaela almost immediately because they wished to be filmed with my parents examining the skull.

This visit proved to be of major significance because it led to my father being introduced to the National Geographic Society. Although my parents knew Michaela and Armand Denis they were not particular friends. Nevertheless, in my parents' camp that Sunday evening, after dinner, a discussion developed about the future work at Olduvai and my father described the great problems of getting financial support for his work. Armand Denis spoke up and proposed that he arrange a meeting with Melville Bell Grosvenor, then the President and Editor of that most famous of American institutions, the National Geographic Society. Armand knew Melville Grosvenor as a friend and he was quite certain that the Society would be interested. My father therefore went to the U.S.A. for the first time in 1960. Armand was right; Melville Grosvenor was interested in the project and apparently liked my father. This led to the initial funding from the National Geographic Society and ever since then the Society has financed work at Olduvai and also at a wide variety of other East African sites. Initially the amounts were quite small but it enabled my father to buy more equipment and further his work. What my parents subsequently achieved must be directly related to the support of the National Geographic Society; their money made possible important finds to which they gave world-wide publicity. The role of Armand and Michaela should also not be forgotten; it was their introduction which launched the relationship.

We bade farewell to my parents at Olduvai and took off on our safari across the Serengeti. After several days, Des had enough footage of Armand and Michaela driving through Africa against a backdrop of wild animals, so they were put on a Nairobi-bound flight. We stayed on and spent two weeks in the western Serengeti where in September, hundreds of thousands of wildebeest and zebra concentrate every year and compete for the limited water at the scattered water holes. Lions and other predators abounded and it really was the romantic Africa – harsh, violent, exotic and filled with vibrant life. The migration of game on the Serengeti plains has to be one of the most remarkable spectacles in the natural world. In December and January, and sometimes into February and March depending upon the rainfall, the plains around Olduvai turn green as new grass grows and small patches of wild flowers and herbs emerge. The ground underfoot is soft and the air is clear, and through this remarkable landscape move hundreds of thousands of animals. Around Olduvai the animals are often milling about for a month or so and sometimes they calve while in the area. It is possible to drive for hours through the herds accompanied by the haunting barks of the zebra and the grunts of the wildebeest. The scene is extraordinarily evocative. This was my first experience of it – previously I had gone no further than Olduvai and my visits had not coincided with the migration. Since then I have seen the migration many times, but it remains an extraordinary experience – something that words cannot adequately express.

I had two amusing lion encounters on this trip. On one occasion, we called on the park warden, Miles Turner, who had a house at a place called Banagi. The house was surrounded by a wide open verandah and although the house was large, it happened that all the bedrooms were full. Miles asked me whether I would mind having my bed on the verandah. Not wishing to appear in any way afraid, I readily accepted, despite the fact that not long before, Miles had been telling us about two large male lions that frequently came on to the verandah during the night. I convinced myself that the lions would not do this if it were obvious that I was there; also, I vowed not to sleep!

Of course, my best intentions failed and by midnight I had been asleep for at least an hour. Something woke me and instantly, appalled by my slackness in falling asleep, I was wide awake. I peered down the length of the verandah and to my absolute horror I could make out quite clearly the forms of two large, maned lions at the far end. My bed was near the front door and so I quietly slipped from under my mosquito net and dashed trembling into the house. I took up watch and after about an hour both lions moved away and I could hear them grunting as they set off, presumably discussing the very unfriendly behaviour of a certain young boy. My dilemma was whether to go back to bed or to sleep on the settee for the remainder of the night. As my greatest concern was not to be discovered by my hero the warden to whom I had given such nonchalant assurances of my bravery I decided to go back out to my bed and to pretend that I had never taken fright. Miles and the other adults were wonderful; nobody questioned me and I was able casually to mention at breakfast that I had seen the two lions but that they had gone off. Little did I know that both Miles and Des had heard the lions approaching while I slept and had kept me under constant surveillance, being ready to intervene had the lions moved towards me. They were kind enough not to hurt my pride by revealing that my flight had been witnessed and ever since I have been very aware of the importance of letting people save face.

The other incident was my first experience of being alone in a small tent with lions moving about just outside. I woke one night to hear coarse breathing and so I shone my torch through the little round window of the tent. I found myself looking at a lioness about five feet away and here again I found myself in a difficult predicament. Should I call out to Des in a nearby tent or should I keep quiet? Pride led me to the latter course but my strong sense of compromise made me carefully turn my camp bed upside down and spend the rest of the night under my bed and bedding! How silly I felt when next morning, Des told me that he had watched the lioness in the starlight and that moments after I shone my torch at her, she had gone off and not returned.

At the end of the filming in the Serengeti we returned via Olduvai, where my parents were in the process of closing down the work until the next year. They were well pleased and confident that at last they would be able to get the financial support they needed to continue their work. This proved to be so and from 1960 Olduvai became a cornerstone in our family activities. Reliable vehicles were obtained along with sufficient supplies of fuel and provisions. A permanent camp was built in 1963 and from then on there have always been people at the gorge to look after the fossil sites and caretake the thatched huts and other camp buildings.

In January 1961 my father encouraged me to make a decision. If I went on at school, he would support me, but if I chose to leave I would have to become self-supporting as soon as possible. I shall always be grateful to my parents for their willingness to allow me to make my own decisions about the important things in my life.

As a seventeen-year-old, although I had no idea what to do for a career, I thought I knew what I did not want to do. There was no doubt in my mind that I should avoid at all costs an academic life and, in particular, I was determined to distance myself from my parents and their work on fossils and prehistory, largely because I wanted to be my own man. I stopped going with my parents on digs at about the time that I left school. Previously, I had gone with them either when they could not leave me or when I could be useful. An outdoor life had considerable appeal and I suppose that like many others I saw a certain romance in being identified as a wild, independent, outdoor young man, although I was still living at home.

I was very lucky when fate gave me just such an opportunity. At the end of 1960 and the beginning of 1961 Kenya and Tanzania were suffering from the effects of a long drought and animals were dying in large numbers. As a result, numerous corpses dotted the plains and the scavengers were so well fed that they left many carcasses quite intact. The vultures and hyenas usually only took the tastiest morsels, such as the eyes and viscera, leaving the rest untouched. I borrowed £500, bought an old Land-Rover and took the opportunity to collect skeletons, which I then sold for about £20 each to universities and museums around the world.

I used to prepare the skeletons by boiling the carcasses in a large steel drum which made it easy to clean away the soft tissues by hand. The bones were bleached with hydrogen peroxide and after sun drying, they were carefully labelled. The work taught me a great deal about the comparative shapes of bones of different mammals and I was not long in becoming quite proficient at identifying the various species on the basis of individual bones, although at the time I had no inkling of how important this skill was to become only a few years later.

In addition to collecting skeletons, I kept up my trapping and I was able to use my father's overseas contacts to good effect. I began supplying several institutions with rare primates and at a good price. One animal I was asked to catch was the potto, a strange primate that is found in greater numbers in West Africa, but which occurs in one small patch of forest in western Kenya. This, as well as the occurrence in the same area of various other West African animal and plant species, is a dramatic reminder that, not too long ago, there was a continuous belt of tropical forest right across the centre of Africa from coast to coast. The potto is a prosimian: a primitive, tree-dwelling, nocturnal primate. Pottos feed on berries, insects and resins, and because such food is plentiful in the lush forest, it is quite difficult to persuade one to enter a trap. My solution was to climb into the tree and catch the potto by hand. This may sound more dramatic than it actually is. At night, pottos are easy to see with the aid of a bright light because their eyes glow a deep red. By walking along the forest paths and shining a torch at the trees it was easy to pick out the glowing eyes and to decide whether the animal was in a tree that could be quickly climbed.

If the tree was suitable – not more than forty feet high and with a slender trunk and plenty of handholds – I would frantically climb towards the potto, while an associate held the potto's attention in a beam of light. Fortunately, pottos are incredibly slow both in mind and limb and the animal's usual reaction was to work its way carefully to the end of a branch. Presumably this is a good defence against heavy predators, but it made my job easier. It enabled me to cut off the branch and lower the animal on its perch by means of a rope, to my associates waiting below. What could be more simple? Each episode took about an hour, but in the process I regret that I did spoil a number of trees. In all, I caught and exported about thirty pottos, which were used to establish a breeding colony in America. All the animals arrived safely and before long their fecundity put me out of the potto business!

There are few animals that have 'treed' me but I was put to flight one night whilst looking for pottos. It was raining and I was leading my party along a slippery track at about three o'clock in the morning. We were all tired and cold having failed in two capture attempts because the trees were so wet that nobody could climb them safely. As we walked we played our torches over the tree tops and made only periodic glances at the road which was made visible by the glow from a kerosene lantern carried for that purpose. On one of my glances at the road I saw a pair of bright eyes some thirty yards ahead and before I could say anything, the animal began to advance. We saw to our horror that it was a ratel or honey badger – one of the most fearless little carnivores known. They have a reputation that may or may not be based on fact; they are said to attack by attempting to emasculate their human foes. Well, slippery or not, we broke all the tree-climbing records that night!

Another primate that I caught for export was the bush baby, of which several species are to be found in East Africa. The most common and so least profitable was the Greater Bush Baby. These are easily tempted by a ripe banana and will often go into a box trap. The main problem was that the trap was often sprung by the animals bouncing on top of the trap as they landed after a leap from a nearby branch. The challenge was to place the trap in such a way as to discourage such frivolity.

Alternatively, I made use of the bush baby's widely known taste for alcohol. By placing a banana saturated in alcohol at the base of a favourite tree, one could sometimes attract several animals which would be found next morning sleeping soundly on the ground. Unfortunately, many bush babies seemed to recognize their capacity for alcohol and would go off to safety before passing out.

One particularly memorable excursion was in pursuit of a much rarer species of bush baby that we were told lived in the forests on the summit of the Uluguru Mountains in eastern Tanzania (then Tanganyika). This was my first chance to lead my own team to a remote place where I would have to live on my wits. I had to get special permission from the Tanganyika colonial authorities to go to the area, and I was thrilled when I was told I could proceed to the mountains, provided that I took along an armed game ranger to protect me. The source of the danger was not fully explained but it all seemed terribly exciting.

I established my base in an old disused forester's house built by the German Administration between the wars, in the foothills of the mountains. I was anxious eventually to make a camp at the top of the 8,000-foot massif, where I believed my bush babies were to be found. Before hiring porters to carry the equipment and supplies up the narrow winding foot trail, I spent several days locating the best route and trying to find a guide who would lead us. In this I failed; the few people we contacted at the foot of the mountain could not be persuaded to go up. I was told that very unfriendly people of a different tribe inhabited the mountain and they would certainly kill us. On my own excursions up it, I had failed to make

ABOVE With my mother, Philip, and my half sister Priscilla, who was visiting us with Justin, her husband

BELOW Proudly mounted on my pony, Susie

LEFT In spite of losing my stirrups, I very rarely fell off

A sequence of pictures showing the release of a captured lion

ROYAL
PARKS

ROYAL
PARKS

My camp in the Usambara Mountains in northeast
Tanzania, where I was trapping animals

any contact with these people although I often heard them ahead, but they always fled into the forest before we could meet. It was a strange and some ways delightful experience; at last I was a brave, intrepid explorer!

I am sorry to say that my heroics did not last long. My colleagues, fellow Kenyans, became very worried and soon had convinced themselves that we would surely be killed, either by spears or, worse, by a magic spell cast by those mysterious mountain people. The intensity of their conviction made our camp a suspicious and unhappy place and I had to agree to give up my plans and to leave. On the way back we called in at a Roman Catholic mission near the main road to Uluguru and we were told that in fact, just a few months before, two young Europeans had been speared to death on their way up the mountain where they were to conduct a mapping survey.

I returned to Nairobi without much to show for my exploits in 'Darkest Africa' and I never did establish whether my rare bush baby, the Needle-clawed Galago, did in fact reside there. On the journey back to Nairobi and while travelling at night we picked out in our headlights the eyes of a bush baby in the dry thorn bush country near Dodoma. I stopped the vehicle and we tried to catch it. As we approached, it seemed to fly from bush to bush, covering distances that suggested the tiny animal had wings. My colleagues decided that this was magic and refused to have any further hand in the hunt. All I could do was satisfy myself that this was the tiny primates, known as the Lesser Galago (*Galago senegalensis*) which has the most amazing leaping ability, covering distances of about twelve feet in a single jump. For a moment, it was hard to believe that the little animal did not have wings and I almost convinced myself that I had discovered a new species.

In addition to collecting primates of the nocturnal variety, I spent some time catching baboons. One of the more interesting requests I had was for a dozen infant baboons for a particular research project in the U.S.A. It was stressed that the animals had to be less than four weeks old and quite free of disease. I had no idea how I was to catch such young animals and I soon found that it was almost impossible to trap them. Infants are held closely by their mothers, and mothers with young will seldom venture into a trap. I eventually decided that the only way to catch such young baboons was to snatch them from their mothers. I had two methods for doing this – both dangerous and neither recommended to would-be baboon hunters. The first method involved chasing a baboon troop by car until there were signs of individuals tiring. I then abandoned the vehicle and took up the chase on foot. I singled out a mother carrying an infant and without too much difficulty, I found I could overtake and grab the fleeing baboon. The vehicle chase was exciting in itself: hurtling across the plains after my quarry required cross-country speeds of up to thirty miles per hour. You can imagine the effect of holes, rocks and washouts hidden from sight in the grass. The one thing I did learn was a great deal about repairing Land-Rovers!

One particular asset I had in my baboon hunt was my dog. I had acquired a puppy from the dog pound the previous year and I kept this animal until he died in 1974. He was called Ben and his pedigree was rather mixed; he was said to have had an Alsatian father and a Labrador mother. Ben became a very special friend to me and I have the fondest memories of this remarkable dog. He especially enjoyed the baboon chase and he was tremendously helpful. His greater speed enabled him to catch up with the fleeing mother more quickly than I could and he would skilfully hold her down by pressure on the neck. Never once did his teeth even break the skin of the baboon. I would arrive, grab the baby and we would run off back to the car before the troop had time to regroup and attack. Baboons are very dangerous animals and it is best not to stay to face an angry troop.

My second method was more dramatic and resulted from my once having watched a leopard hunting baboons. The leopard had come across a troop that was settled in for the night high in the trees. Baboons usually take up their roost at dusk. On this occasion, the leopard managed to get under an acacia tree without detection, but once there, made no effort to conceal itself and indeed seemed anxious to be seen. To my surprise, a large number of baboons came leaping down; they had panicked and instead of staying in the safety of the tree they had jumped for their lives! The leopard was able to pick off the animal of its choice without any effort.

I slightly altered the leopard's technique because I could never get under the tree at dusk without being seen. My way was to creep in before dawn and to wait as the first light began to colour the horizon. As soon as it was light enough to be sure of making out mothers with babies, I would pretend to be a leopard, growling and making the characteristic hunting sound of this proficient baboon hunter. Sure enough, down would come a number of baboons and it was easy to grab an animal. To give credit to the baboons, mothers with young generally stayed in the tree and it proved a difficult way to catch babies. As in the case of the chase we had to be quick because it was not long before the troop regrouped and attempted to rescue the screaming captives. The other problem was that baboons have the unfortunate habit of emptying their bowels in the early morning hours; sitting below forty or so such animals for half an hour required considerable self-discipline!

Despite the difficulties, I was successful enough to have had a number of animals in cages in our camp when one day a youngster escaped and rushed into a large tree that was shading the tents. I thought I would use my potto technique to recapture it but that method did not work. Nor could I get the animal to jump out of the tree so, after many fruitless hours, I devised what I thought would be a better plan. The tree had grown in such a way that it was impossible for the baboon to move from one side to the other without using certain branches. My plan was to put someone in reach of this 'passage' and to give the baboon a fright by shooting into the tree in the hope of making it race past an assistant who would then be able to grab it.

I asked a young man who was working with us to climb the tree and act as catcher but, apparently, I failed fully to explain my plans. As soon as he was in position I asked him if he was ready and upon hearing a muffled 'Yes', I fired a shot into the tree, close to the baboon but at least twenty feet from the young man. To my absolute horror and surprise, he gave out a cry and fell to the ground. Could I have missed my target by twenty feet and shot him? I and my associates rushed over and found him to be quite alive but shocked. As soon as he too realized that he was unhurt, he jumped to his feet and bounded away not to be seen again for several days. Muteti, as he was called, had for some reason that was never explained, suddenly thought that he was to be shot. Upon hearing the report, he reacted accordingly. When he found that I had missed, he bolted and returned to his village nearby. It took some persuasion to convince him of what actually happened and I am happy to say that he worked for me regularly thereafter. We never did catch the baboon, which apparently came down the next day when we were all away and joined a troop that came by the camp.

There were various other animals that I trapped or caught for one reason or another and my youth was very much animal oriented. As children we always had wild animals as pets and my parents' house with its central grass-covered courtyard was ideal. Our pets included a variety of duikers, bush babies, bat-eared foxes, a baboon, pottos, genet cats, an eland and a wildebeest. My mother had a hyrax, and still does to this day. My brother Jonathan had an Eagle Owl which was quite free but chose to live in the house. This magnificent bird we called Ferdinand and it was with us for at least four years. He would often bring rats and mice home to share with us and on many occasions I awoke to find Ferdinand perched on my head trying to force an almost dead mouse into my mouth. This was much more of a problem when my parents had guests who left their windows open at night!

After leaving school I worked on occasions as a general camp assistant at Olduvai. This gave me valuable lessons in how to organize supplies and to manage people. The time at Olduvai was idyllic and I saw many wonderful sights there. Once, in 1962, I was driving a Land-Rover to Ngorongoro to collect water for the camp. The vehicle was filled with drums and I was trailing a 500-gallon tank or water-trailer. All the water requirements of the Olduvai camp had to be fetched in this way and the task used to occupy a full day. I was on my way up the slope of the crater, the vehicle grinding slowly up the steep incline when suddenly a lion raced across the road about a hundred yards ahead. The lion was obviously in a hurry, but before I could give this any thought I had another surprise as several Masai warriors also raced across the road quite oblivious of the oncoming vehicle.

I realized that I was witnessing a lion hunt, a sport for which the Masai were famous before the introduction of strict laws for the protection of wildlife. I decided to abandon my car and trailer and I raced off in hot pursuit. I was determined to witness the dramatic spectacle about which I had heard so much. It was said that one of the warriors grabbed hold of the lion's tail while the others speared it to death. Fortunately the lion was tired and I did not have to run far to catch up. I arrived on the scene to find a group of about twenty warriors surrounding the angry lion, their spears at the ready to stop any attempt to escape. They were gradually drawing the circle tighter. I kept a respectful distance, not wishing to be a distraction nor wishing to become an unarmed participant.

As I watched the chanting and frenzied circle of warriors, one of them thrust his spear in the ground and raced in to grab the lion's tail. But he was too slow, the lion turned and swatted the man with a front paw, killing him instantly. This increased the amazing tempo of excitement and a second warrior rushed in and managed to get hold of the tail briefly. However, the lion shook him off by a violent turn and in the process managed to maul severely the man's leg and groin. A third warrior raced in, grabbed hold of the tail and although the lion was furiously biting at his upper arm and shoulder, he held on grimly while the others quickly speared the lion to death. The whole incident probably lasted no more than two minutes but it seemed an eternity. The excitement was such that several of the warriors passed out, frothing at the mouth and with wide staring eyes. Others, singing and dancing, dragged the two wounded men to the shade of a thorn bush and quickly applied packs of dry earth, leaves and grass to plug the holes and stem the bleeding. The wounded warriors, quite oblivious of any pain, joined in the singing and dancing with amazing vigour. The man who had held on while the lion was speared was given the tail as a trophy while the others hacked off various bits so that back at their village they could prove their own prowess and the part they had played in the hunt.

My presence was hardly noticed while all this was going on and it must have been at least ten minutes before a senior warrior came over to where I stood and asked me to take the two wounded men to the clinic at Ngorongoro. I was fairly well known by the local Masai and I was always well received and treated as a friend.

I witnessed another amazing example of the courage of the Masai and their unwillingness to admit pain when I took a man to the Ngorongoro clinic one day. I had received an urgent summons for help and, on inquiry learned that a man had been speared while defending his cattle in a night raid by another band of warriors. We drove across country for about seven miles until we reached the village where the wounded man was lying. A spear had been thrust through his abdomen just below the ribs some six hours previously. He was in remarkably cheerful form despite a tremendous loss of blood and a ragged wound. The cross-country drive followed by a two-hour ride along the rough road to the clinic was taken without any complaint – in fact, a lot of time was given to a discussion with his friend about the revenge that would follow. Whether another raid occurred I do not know but this was the essence of life at that time for the pastoral Masai of the Serengeti and it greatly impressed me; they were a brave, fearless people who knew nothing of luxury.

CHAPTER FIVE

SAFARI BUSINESS

As SOON AS THE NATIONAL GEOGRAPHIC SOCIETY had agreed to support my parents' work, the President and Editor, together with several officers, felt it to be essential that they visit Olduvai to see the project. So it was that I got the chance to act as a guide. Not only did I help at my parents' camp but also further afield and I managed to persuade a number of these Geographic friends to tour East Africa. In due course I began to offer myself as a regular safari guide, using my own Land-Rover, and charging fees for the service.

In 1962, I decided to establish a commercial company that would formalize my increasing activity as a safari guide. As a result of father's publicity, I had many excellent contacts and, at the age of eighteen, I envisaged myself as a very successful company director, raking in money from tourists who were anxious to have a 'special trip'. I played heavily on my father's reputation and in fact made a good start. I invited a friend, the brother of my then current girl friend, to join me in the venture. This was probably a mistake. Our association never really worked too well because there just was not enough business for two and the tourists wanted to be guided by Louis Leakey's son. Of course, I capitalized on this because otherwise it was hardly likely that people would have wanted a youth like me to lead them on expeditions into the African bush.

One of my first major safaris was in 1962. I was commissioned to organize a camp at Olduvai for a television team for the National Geographic Society who were to film the work of my parents. The job involved buying tents, extra vehicles and all sorts of equipment, and thereafter providing my 'clients' with the full comfort of a luxury safari camp in Africa. I was able to earn a lot of money from this project and, of course, the extra vehicles and equipment proved essential for our continuing safari business.

During the course of the TV safari, lions frequently came into the camp and I had tremendous fun showing off my confidence with lions. If only my Geographic friends had known how scant this was! Each morning at breakfast, after a night of lions in camp, I would enjoy teasing various members of the party about their obviously sleepless night. One morning I was particularly obnoxious and unprofessional, but later that day my clients got their revenge. They had chartered a tiny fabric-covered aeroplane to enable their photographer to get some slow-flying aerial shots of Olduvai Gorge. I had been conditioned by my parents to distrust and fear small planes and this piece of information was somehow obtained by my friends who casually offered me a flight. I tried vainly to be too busy but there was no choice, my honour was at stake. The pilot had been fully briefed; he gave me a terrible flight including steep turns, low flying and quick levelling out after sharp climbs. I was terrified. My colour and expression upon landing fully rewarded my friends for all their suffering at breakfast!

The consequence of this flight was that I became convinced that I had to overcome my fear of planes and the obvious way was for me to learn to fly. I knew that my parents would not allow me to do this but as they planned to be away in March 1963 on a trip to America I arranged to have my first lesson a few days after they left. To do this I cheated a little on my age; I registered myself as being ten months older than I actually was. By the time my parents returned I had had my first solo flight.

On one occasion while I was working as a safari guide, I had a camp near the Mara River, at a spot which today is in the famous Mara Game Reserve. At that time the region was protected but there were no lodges and anyone visiting the area had to be self-sufficient, taking along tents, beds, bedding and their own supplies of food. It is beautiful country now, but in those days it was even more spectacular because it was so far from the beaten track. There were no cars and no bus loads of tourists taking photographs of every animal that might be about in the daylight hours.

I had a small group of clients and we had just sat down to an early afternoon tea before setting out to see game. The table was set under a huge tree in front of the tents and while I was pouring the tea some impala antelope came bounding through the camp. This was very unexpected because wild animals seldom come into a camp in daylight. The next thing we saw was a wild dog rushing past in pursuit of the impala and I immediately realized that we were witnessing a wild-dog chase. Within moments the dog had disappeared, only to reappear, this time pursuing a half-grown impala that came right into the camp and stood at bay close to where we were sitting. The young animal showed no fear towards us. Nor did the dog. Suddenly, three more rushed in and seized the impala by one of its hind limbs. Without thinking, I also grabbed the impala and there was a tug of war between the four wild dogs and myself as to who was to gain possession of this unfortunate young animal. Several more dogs appeared on the scene and I began to realize that it was extremely foolish to be doing what I was. My clients had taken refuge in one of the Land-Rovers and my staff had climbed into the back of the truck. I decided to let go of the impala lest the dogs decide that I too should be included in their meal. In a matter of minutes the impala had been torn to shreds, devoured, and the dogs had moved off to hunt elsewhere.

Wild dogs very seldom attack people although they have always been a serious pest to farmers because they rampage on the farms destroying large numbers of goats, sheep and cattle. As a result, wild dogs are fairly rare in Africa today, hundreds of thousands of them having been killed over the last fifty or sixty years. However, there was one incident that I remember when dogs did attack a human.

It was in 1960 when I was at Olduvai with my parents, who had an unusually large camp because of the new scale of excavations following the *Zinjanthropus* finds and the receipt of the National Geographic Society's grant. In those days there used to be a very rough track across the Serengeti Plains to a little village known as Loliondo some one hundred miles north of Olduvai. This track was used only once or twice a month, generally by traders who went to the village with supplies which they traded with the Masai. A truck, driven and owned by a Sikh, who had been trading this route since World War 2, broke down during one of his trips. The point at which this happened was probably closer to Olduvai than to the village and in the absence of any prospect of another vehicle coming their way, the two occupants decided to walk for help.

They set out for Olduvai where they knew of my parents' camp but on the way they were attacked by a pack of wild dogs. There was very little opportunity to escape because the plains are wide open and treeless. Each man took to his heels running in separate directions thus providing a distraction to the pack of dogs. One of the men eventually reached the Olduvai camp and sought assistance from my father who immediately sent off a vehicle to look for the other man. The search resulted in a gruesome discovery. Two shoes, a turban and some garments were found scattered over the area where the attack had occurred. The man had been torn apart and totally devoured.

In 1964, I was leading a safari that entailed spending some days on the Serengeti Plains. One of my camps was established at the north-western end of Olduvai Gorge where there is often a shallow soda lake called Lgarga (now known as Ndutu). In the dry weather there are sometimes considerable numbers of flamingos which feed on the abundant algae. It was a wonderful camp site because it overlooked the lake and, in addition to flamingos, there were countless other species of birds to be seen. I had

In the Ituri forest, Eastern Zaire, with some Pigmies

ABOVE Supervising the loading of our equipment for the 1964 Natron expedition

BELOW Our raft could be used in the shallow water of the lake

on my safari two couples and one of the men was an extremely enthusiastic bird photographer, for whom I intended to construct a hide on the water's edge from which he could photograph. There is now a well-organized tourist camp at my old camp site, but in those days it was deserted and there were no roads into the area.

The first evening in camp was exciting because a pride of lions came by and their roaring caused quite a disturbance. This was an essential ingredient to the atmosphere of any luxury tented safari and everybody had a marvellous time talking about his or her personal experiences and reactions to the lions prowling around the tents at night. During the following day I set up the hide. The photographer and I soon discovered that the flamingos would fly away if they saw us walking out towards them, but that when we left the hide they would, after a period of time, come back. The problem was our impatience: none of us had the time to sit in the hide long enough for the birds to become used to us. We therefore decided to leave the hide for a whole day with a view to occupying it in the early hours of the morning under the cover of darkness. In this way I hoped that the flamingos would not see us arrive and think that the hide was as it had been the day before, without people.

During the second evening, we again had lions in camp and I was a little concerned about the prospects of getting up and walking the half mile to our hide. I hoped that by five o'clock in the morning they would have moved off. I woke my photographer friend at half-past four and while we were having tea before departing for our hide, he asked me about the lions and I replied to the effect that I was certain that they had departed. As we set off walking across the salt flats of the lake, I became acutely aware that we were not alone. I could hear footsteps some distance behind us crunching on the soda crust which covered the ground. I realized that we were being followed, almost certainly by the lions. I hoped that my companion would not recognize the noise of footsteps and that we would be able to carry on and reach the hide. In retrospect it was probably a foolhardy decision and I might have done better to have abandoned the whole project.

We were about half the distance across the flats when my companion heard the noise, tapped me on the shoulder and said 'Something is following us; there is something behind us.' There was, of course, no doubt about it but for his peace of mind I told him that the noise was being made by curious hyenas, which were no cause for worry. Although I do not think I convinced him we continued to our hide, where we were able to get behind the flimsy screen. The hide consisted of four poles around which I had tied some burlap sacking. The whole thing was about four feet square, the sacking stood at something less than five feet high and I had cut a little hole through which my client could take his photographs. We sat in the hide for the next twenty minutes waiting for the dawn to break. I could not hear the lions any more, there did not seem to be any movement and the only sound was the characteristic noise of the flamingos, chattering to themselves as they fed in the shallows some thirty yards from us.

As the sky lightened, it was clear that we were going to have a magnificent dawn and my companion got more and more excited at the prospect of taking his photographs. When more light advanced across the sky I peered over the top of the hide and saw several shapes but as I could not remember any irregularities on the surface of the lake around where we were I became a little suspicious. In the increased light I realized that the shapes were, in fact, three lions sitting about thirty feet away, watching us with great interest. They obviously found it extremely amusing to see these two strange people wander out across the salt flats, climb into this tiny contraption and sit there in absolute silence.

As the sun came up and it became light enough to take photographs the lions moved off towards the far side of the lake. It was with a great sense of relief that we saw them go and we turned our attention to flamingo photography. No sooner had my friend started his preparations than I observed another lion making its way around the edge of the lake towards us. On this occasion it was a large male and it was clear from the direction in which he was looking that his first thought was to investigate this strange object set out on the edge of the water. I regret to say that at this stage I decided I had had enough. I pointed out to my colleague the lion which was coming our way and we both decided that the most sensible thing would be to abandon our hide as fast as we could and scurry back to the camp. He never did get his flamingo photographs.

The safari business gave me a wonderful chance to explore East Africa's wildlife areas and I took great delight in planning trips to the remotest places for my clients. I was tremendously happy for a while but I still wanted something else, although I didn't know what. I began to find that the paying guests were in fact often an obstacle: they, naturally, wanted to keep moving so as to see as much as they could, while I, having seen it before, often preferred to linger. I began to spend more time at Olduvai, where I still worked as camp assistant, than actually earning my living as a safari guide.

At about this time, I decided that my safari business could profitably be expanded. One of the things that particularly interested me was the idea of establishing a tree house, rather like the famous Tree Tops Hotel on the slopes of Mount Kenya. Tree Tops Hotel is simply a large tree house. Originally there was a small game-viewing platform which grew into a more substantial affair. In 1951, the present Queen Elizabeth and her husband Prince Philip were staying at Tree Tops when they learned of the death of King George VI. Subsequently the tree house was burned down and the present lodge is built in the same place but supported by posts. From the lodge it is possible to watch animals come in to drink at the waterhole without disturbing them. The site I had selected was on the rim of Ngorongoro Crater in Tanzania. I had found a natural salt lick in the forest adjacent to a pool of water, and the view from the tree was across the Crater to the highlands east of Ngorongoro and to the Rift Valley beyond. On a clear day one could actually see Mount Kilimanjaro which was at least 150 miles away. My first concern was to establish whether I could attract a regular supply of animals to the site. I scattered salt around the pool and in a very short time there was a noticeable increase in the number of animals visiting the area. I became increasingly confident that it would be possible to establish a very successful competitor to Tree Tops.

The idea of building a tree house as a tourist lodge was rather daunting so I decided to discuss the matter with a number of different people who might have some useful advice. The first thing that became clear was that the tree house could only be run successfully if it were affiliated with a conventional hotel where services such as laundry and basic catering could be taken care of. The house itself should only be used for overnight visits as it was important to keep the disturbance in the immediate vicinity to a minimum so as to encourage the maximum amount of wildlife activity. Being fairly naive about such matters, it seemed to me the only thing to do would be to buy the existing lodge – then known as the Crater Lodge, Ngorongoro – which had been successfully operating for at least fifteen years. With the assistance of my friend Bob Campbell, who was a photographer, I began to investigate my chances of raising the necessary capital. I also had to persuade the Tanzanian authorities to provide us with the necessary permits, so I flew to Dar es Salaam with Bob to meet representatives of the government department concerned with tourism. It soon became apparent that while they were quite willing to encourage us to go ahead, they had some reservations. It turned out that the Tanzanian Government was planning to nationalize all tourist enterprises. I have never had so narrow an escape in any of my business ventures before or since!

It was during this period of my life that I was invited to stand in as the safari guide or 'white hunter' at Tree Tops Hotel. The tree house then accommodated about fifty people; the regular hunter had fallen sick and I was very pleased to earn some money although more than aware that I was not in any sense a 'hunter'. My duties began at the Outspan Hotel at Nyeri, where my party was having lunch before leaving for Tree Tops. I had about forty-five people in the group, mostly middle-aged to elderly and of varied nationalities and interests.

We set off in a convoy and I took the lead vehicle. When we eventually reached the Tree Tops car park, I told everyone to remain in the vehicles while I went ahead to check that the path was clear of dangerous animals. I did not expect for a minute to find any – it was partly a precaution and partly to build up excitement for the expedition. I set off up the footpath with my rifle casually over my shoulder, but I had gone less than half the distance when I heard elephants. I went on rather less enthusiastically and found to my horror that there was a herd of about forty all around the tree house; a good number of them seemed to be feeding on both sides of the path between where I was and where I had to take the hotel guests. Nobody had told me what to do in a situation like this!

I watched for a few moments and then went back to the cars. I decided that the first thing to do was to get all the vehicles turned round to face the way we had come. In the event of difficulties this at least would ensure a rapid means of escape. I told my visitors what was going on although I think that most did not really believe me. I then returned up the footpath and found to my dismay that the elephants had not moved at all.

By this time at least twenty minutes had elapsed and my visitors were getting a little impatient. They had paid good money to go to the Tree House, not to sit in a car on the edge of the forest with nothing to see. I decided the only thing to do would be to try to frighten the elephants from the path. Again I went back to the tourists and told them that I planned to move the elephants by shooting into the air but that they should not be alarmed.

Once more I went off up the path and got as close as I dared to the elephants, ensuring that I was near a tree with a ladder before I fired the rifle. It was a very heavy-gauge gun that I had been loaned for the occasion and as I fired I was very nearly knocked off my feet. To my horror a cow elephant came rushing down the path towards me in a light-hearted charge. I fled back to where my party was, arriving rather breathless and very red in the face. Fortunately the shot did have some effect and within a relatively short time the elephants moved off and I was able to get everyone into the tree house. We had a marvellous evening; a lot of game came and everybody returned safely the following morning. But it was the last time that I offered to be the hunter at Tree Tops: the whole experience had been unnerving and very embarrassing.

CHAPTER SIX

MAKING MY MARK

IN 1963 WHEN KENYA finally was granted political independence from Britain I was young and too preoccupied with my own life to take much interest. My parents were busy at Olduvai, and at the time were seriously considering moving permanently to Tanganyika (now Tanzania). By September, I had obtained my Private Pilot's licence and I took my first flight as a licensed captain-in-charge to Olduvai. On the way, I saw some exposed sedimentary deposits along the western shore of Lake Natron. These fascinated me because they looked very similar to those at Olduvai, although on a much smaller scale.

So curious was I that I resolved to organize a short expedition to travel overland to see whether these stratified silts and clays actually had fossils preserved in them. My parents were in favour and agreed to loan me a vehicle and some of their staff. Glynn Isaac joined me, as did another friend, and we had a wonderful fortnight exploring the western shores of Lake Natron.

We did finally achieve our objective and were delighted to find that, just as we had hoped, there were fossils to be found. We called the site Peninj after the river by the same name that flows through the area. The potential for further work and, of particular importance to me, the chance for further adventure was clear and we began to plan for a longer visit. So it was that, with financial support from my father's Geographic grant, Glynn and I returned in early 1964. On this occasion I decided that it would be easier to approach from the north, use a boat to cross Lake Natron, and then walk to Peninj. I had some wooden pontoons made in Nairobi with which we constructed a shallow-draught raft to cross the six or seven miles with all our equipment.

When we eventually arrived on the shore of Lake Natron, I was particularly pleased to find that the lake level was very high. I had thought that we might be unable to get across the lake by raft because of the excessively shallow water and mudbanks, but the rains had brought Lake Natron to an unprecedented high level. There was a continuous stretch of water from the north end of the lake right into Tanzania to the base of the active volcano, Ol Donyo Lengai. One of the consequences of so much water was that the swamp at the north end, where the Ewaso Nyire river drains into the lake, was also full of water with a very large stand of swampy vegetation.

My plan had been to establish a base camp on the shore from which we could get everything across to Peninj. Although the camp was by no means permanent, some people were going to have to stay there for the duration of our expedition. It was therefore important to camp near the mouth of the river where we could have a supply of fresh water. Little did I realize the implications of this decision.

On the first day after we had arrived, we made a temporary camp and simply set up beds out in the open; there was no prospect of rain. By six in the evening the wind had dropped and the mosquitoes began to appear. By dusk we realized that our problems were only just beginning. I have never in my life seen quite so many mosquitoes at one time. And never before had I been so badly bitten. My dog, Ben, found it impossible to sleep and he spent the whole night whimpering and howling with pain. The next morning his feet, ears and whole face, particularly around the eyes, were puffy and swollen from the hundreds of bites that he had suffered in the night.

We had to spend a second night at this camp because it took two days to get everything across the lake. The next evening we all retired to bed before the main onslaught of mosquitoes began. Sleep was virtually impossible because a large number of them penetrated our nets. I had Ben under my net with me and he continued to get bitten to such an extent that next day he was barely able to walk because of his swollen feet and legs.

Eventually we had everything across the lake and were ready for the final move to our camp site near the Peninj River. I hired thirty or forty porters from the village to transport all our gear. Once assembled, they took up their loads and set off on the two-mile march. In those days I used to have a particularly strong social conscience, and so I too felt obliged to carry a heavy load. Accordingly, and to show solidarity with the others, I took charge of a cumbersome wooden box weighing at least forty pounds and with it I took up the tail-end of the line. Naturally, with this load, I could not catch up with the head of the column and before long there was chaos. The line broke, the porters pillaged the loads without interruption and then took off. We lost blankets, sheets, pots, pans and a lot of food such as rice, sugar and flour. It was a good lesson. Now I make sure that when I am in command I am in the lead!

My expedition included my younger brother, Philip, as well as Glynn Isaac, now a well-known archaeologist, who at that time was working with my father at the Museum. Quite apart from my friendship with Glynn, my parents felt strongly that I should be accompanied by someone with some academic training and competence. We were later joined by Hugo van Lawick, a good friend and a fine photographer, best known today for his work with Jane Goodall on chimpanzees, who was then without a job. The National Geographic had wanted the expedition to be filmed and I suggested Hugo, who was then no longer filming for Armand Denis. We also had about six Kenyans with us, all labourers who could help with the camp and if necessary with excavations. One of them was Kamoya Kimeu, who was then a junior member of the team but whose company I enjoyed and who has since become one of my closest friends and the main strength behind the organization of my current project at Lake Turkana.

After about ten days in the Peninj area, we had settled down to a routine of work and Glynn had embarked upon a series of geological traverses. We found a few fossils of antelopes and pigs, and some stone tools, such as hand axes, but nothing really to excite me. I had enjoyed the challenge of getting the expedition to the camp and everything set up and I looked forward to the organizational aspect of leaving again, but the fossil hunting and archaeology left me rather bored after such an adventure. Thus I took the first legitimate opportunity to return to Nairobi.

I was accompanied on this trip by Hugo, one or two members of my staff and my dog Ben. The journey across the lake on our raft was uneventful and on arrival we climbed into a Land-Rover and set off to Nairobi, a journey of about five hours. The first part of the trip was through bush country where there were a number of antelopes and other animals. I knew that I had to be careful to keep Ben from jumping from the car, and whenever animals were seen ahead, one of us would hold Ben by the collar. Unfortunately, on one occasion we did not see some antelopes in time and before we knew it Ben was out of the window and off in hot pursuit. I stopped the car hoping that he would be back in a moment.

We waited at least fifteen minutes and when Ben had not returned I realized that he was not going to come back. Either he had run into trouble or else, more likely, he had got lost. The grass in the area was fairly long so the visibility for a dog was poor and the wind was also in the wrong direction. I was heartbroken and determined to do everything I could to find my friend.

We carried out a preliminary search by car, leaving someone at the place where Ben had jumped out, just in case he found his way back to the road. After about an hour it was quite clear that we were not going to find him simply by driving around hooting the car horn, calling and whistling. I decided that the only way to find him would be to carry out an aerial search. I again left a person on the roadside so that if Ben did find his way back he would meet someone he knew, and drove hurriedly to Magadi, a small nearby town in the Rift Valley where there was a telephone. I phoned a charter firm which had done a lot of flying for me in the past, and I asked whether they could let me have an aircraft immediately to fly to Magadi. I was told that unfortunately there would be no aircraft available until the following day. This was bad news; the following day would be too late for Ben. In desperation I explained that a member of my expedition to Lake Natron had got lost and I had to have a plane at once to carry out an emergency search. Under these circumstances, of course, a plane was released. Some unfortunate person did not get their scheduled charter and instead a pilot flew to Magadi where I was picked up.

Once we were airborne, I explained to the pilot where I wanted to go. In a very short time I found Ben who was on the ground making rather slow and painful progress across the rough boulder-strewn country but more or less in the right direction. The pilot was astounded to discover that all we were looking for was a dog! Fortunately, he also had a good sense of humour.

It was while I was away in Nairobi that the excitement began at Lake Natron. Kamoya discovered a complete lower jaw of '*Zinjanthropus*' projecting from a cliff face just a few feet from where I had myself been searching before my trip back to Nairobi. What a moment it was! I received the news from Glynn who reached me via the radio telephone at my girlfriend's home. I contacted my parents at Olduvai by radio the next day, and flew there to collect them before I returned to Peninj. I had asked Glynn to leave the discovery well alone so that my parents could see it prior to its excavation and also to enable the work to be adequately photographed.

There it was, 'my' – or more accurately 'our' – first discovery, and at that time the only known lower jaw of the East African species, *Australopithecus boisei*, previously known only by my mother's find at Olduvai in 1959. It took a few days to excavate the jaw and, once collected, father felt that we should conclude the expedition for the time being but return later in the year having raised additional funds. It was his money and his responsibility – also I was delighted to avoid the prospect of a long drawn-out dig! There was a lot to do and I was back in my element organizing porters, the boat, and all the tiny problems that are such a challenging part of any expedition.

Using Hugo's photographs and film, father persuaded the National Geographic Society to put up funds for a three-month project in the summer of 1964. I was to be the organizer, Glynn the scientist in charge of the research, and my father retained overall charge of the venture, although he did not spend more than a few days on the expedition. The longer term and the nature of our project required that we had vehicles on site and I devised a spectacular way to get them there.

Instead of the tough overland route from the south that we had used in our 1963 reconnaissance, I proposed to transport two Land-Rovers and fuel to Peninj by air. To manage this, I had to persuade the Commanding Officer of the Kenya-based detachment of Britain's Royal Air Force to let us use one of the RAF transport planes. My powers of persuasion must have been quite good because he agreed to put a Blackburn Beverley, a giant four-engined, large-bellied transport plane, to the task. The agreement was that the RAF would take us in, but, at the end of the season I would have to withdraw overland without them. We never really discussed the fact that Peninj was actually in Tanzania rather than in Kenya, but I was reasonably sure that there would be no repercussions because nobody would ever know that the Beverley had crossed the border. I hate to think now of the political crisis had there been an accident! As it happened, we were lucky and everything went according to plan.

The first task was to clear a suitable airstrip at Peninj in an area where the ground was reasonably firm and level. I supervised this myself before returning to Nairobi for the flight. As we made our final approach in the huge plane I began to have serious doubts about my airstrip although the pilot assured me that the strip of cleared land, just 450 yards long, was sufficient. The real problem was that the noise and shadow of this vast aeroplane panicked the local herds of cattle, sheep and goats. The Masai and Wasonjo herdsmen had to spend the rest of the day sorting out the considerable confusion. They were justifiably angry but when we allowed them to climb into the plane and inspect it they were so impressed that we were forgiven.

The west side of Lake Natron was very isolated and there were many difficulties for the local population. The Wasonjo people were agriculturists living in small villages along the Peninj River and they were entirely surrounded by the nomadic pastoral Masai. All these people had a walk of at least eighty miles to the nearest trading store or medical clinic. The Wasonjo were particularly interesting in that they practised a very ancient system for irrigating their crops. The river was dammed, not by making a waterproof barrier to raise the water level, but by creating a sediment trap of logs and vegetation which effectively raised the bottom of the river. This not only raised the water level, but obviated the need for complex engineering. The water was directed to the fields by a series of small channels and furrows which were carefully maintained and controlled. The elaborate system of water distribution and the role of the elders in deciding on water rights was obviously a central feature in the social organization of the small community. We ran a clinic for them and got to know a number of the villagers well, particularly the leaders. They would often give us goats in return for things that we brought them from Nairobi. It was a delight to be able to observe the ways of a people who were clearly very much at home in a very difficult environment.

After several weeks, our clinic, from which we dispensed cough mixture and aspirin and treated malaria cases, was drawing daily several dozen people from miles around. Our patients needed a wide range of treatment. On one occasion when I was away, Glynn was called out to stitch up a large wound that a warrior had received in a fight. Glynn who had never had to sew up a wound before, successfully put in about fourteen sutures and in a short space of time the man fully recovered. On another occasion, we sat up for hours trying to save the life of a child who had pneumonia, but we failed. Even the limited help we provided saved many lives and illustrated for me the value of medical aid in the remote areas of Africa.

With two vehicles, we were able greatly to extend our range and we conducted a large survey of the Pleistocene deposits that outcrop along the western shore of Lake Natron. (The Pleistocene period is between 1.7 million and 40,000 years ago.) These particular deposits have been dated at around 1.4 million years. Glynn and I found several archaeological sites, which were subsequently excavated by a small team under Glynn, and for three months I kept myself busy. The large excavation at Kamoya's *Australopithecus* site was a slow business, and unhappily nothing further was found. We did collect some other fossils, including antelopes and hippos, but the locality was small and fossils were not abundant.

I knew, of course, that I had to face the problem of how to get the vehicles out of the area at the end of the project. Our original overland route along the shore was flooded by the very high water level of the lake that year. The only option we had was to improve and widen an existing cattle track that went up the 1,500-foot escarpment to the west of the lake. Fortunately, in the event, this proved not too difficult and after a few weeks' work, I had a track that could just be negotiated by a four-wheel-drive vehicle.

The 1964 expedition consisted of Glynn, myself and several students, one of whom was a girl whom I had known before as my older brother's girlfriend. Margaret was a student of archaeology at the University of

Edinburgh and she had worked the previous year with my parents at Olduvai. While at Peninj, we became quite close but in September Margaret went off to England to continue her studies and I found myself back in Nairobi wondering what I was going to do next. I had had a thoroughly enjoyable three months at Peninj, spending a lot of time with Glynn digging out fossils – an activity that gave me considerable pleasure. In many ways it was so much more satisfying that conducting tourists round East Africa's National Parks.

I was very conscious of the fact that on the Lake Natron expedition I had been at the bottom of the academic scale and although I was organizing the expedition's supplies, transport, and personnel, I could not make any contribution to the scientific publications that were to follow. Working with my parents as a child had, however, given me a basic knowledge of fossils and stone implements but I realized I still had a lot to learn. I needed, too, to find a basis from which I could justify being sent on more expeditions. The question was, how? I had left school without completing my university entrance requirements and when I inquired at the University of Nairobi, I was told that I had to complete my high school qualifications before even being considered for a place. Naturally, having been independent for several years, I was reluctant even to contemplate returning to school and so I thought I would go to England and try a university there. I hoped they would be more understanding. My safari company was becoming less important to me and I had enough money in the bank to pay my air fare to England and to be able to live there.

In early 1965 I was invited by the National Geographic Society to accompany my father to the States. My father was scheduled to report to the National Geographic Society on the work at Olduvai and Lake Natron. While in England on my way to the U.S.A. I went to Cambridge where I made arrangements to meet a professor in the Archaeology Department. I explained what I wanted and tried to gloss over the fact that I lacked 'A'-levels. Kindly but firmly he told me to complete my schooling before wasting anyone else's time. I was disappointed but I realized that I could live in England and study for the exams which could be taken whenever I felt ready. I wrote to a 'college' that specialized in helping mainly foreign students to get through their 'A'-level exams in the shortest possible time and I arranged to enrol in the spring of 1965. I was very pleased that for once I had made a decision about my future of which my father really approved and, moreover, I would be able to see Margaret fairly often. I duly went to America and returned briefly to Kenya to arrange for someone to run my safari firm during my absence.

In due course, all was arranged and I returned to London. I got to work immediately and soon became engrossed in mastering 'A'-level physics, chemistry and biology, all of which I needed for a science-oriented university training. After six months I sat the exams and did well enough to qualify for a university, probably not Cambridge, but nevertheless the first hurdle had been cleared. Unfortunately, during my absence in London, my safari business had not done very well and it was quite obvious that I had to return to Nairobi to sort things out. There was no chance of getting a place in any university before the autumn of 1966 anyway, so I went home early in that year.

Back in Kenya it was not too difficult to sort out the problems in my safari company and I soon got things going again. One of my first contacts on my return was to prove quite fortuitous for the business. A large American television company was to film in East Africa, and I decided to bid for the contract of providing its logistic support. At that time my company was very small and the only way I could hope to get the contract was to quote a very low price. Because I did not have the money to supply such a large expedition I planned, unknown to the television people, to buy most of the equipment for the trip from their first payment to me. In the final round of negotiations an even lower price was quoted from a quite respectable competitor and I had to expatiate on the size and flexibility of my company. I specifically mentioned the matter of trucks. I explained that my business was so effective that if, for example, one of the large trucks happened to be written off in an accident I had sufficient extra gear and vehicles to replace it in twenty-four hours! What a lie! I did not at that time have a truck at all or even enough tents.

Margaret and I had decided to get married and since this coincided with the time of the safari I handed everything over to my friend Bob Campbell; I hoped that he would benefit from the American TV contacts to be made. Poor Bob, the new lorry that I had purchased and sent off loaded with tents, beds and bedding was involved in an accident on the very first day and had to be written off! By luck none of the equipment was damaged and I persuaded the insurers in Nairobi to give me immediate credit so I could buy a second truck the same day. While the TV team spent the night comfortably in the Manyara Hotel I drove the new truck to the scene of the accident where we transferred all the camping gear and equipment.

I have always been deeply grateful to Bob and in spite of the problems on this expedition we have always remained close friends.

I realized that my safari business had to be more soundly organized;

relying on friends such as Bob was not an adequate solution. As a result I took as a partner another friend, Alan Root, who had also been involved with safari tours and so was born Root and Leakey's Photographic Safaris Ltd. Alan Root, now well established as one of the world's finest wildlife photographers, had been a friend since I had helped Des Bartlett in my school days. Alan's increasing preoccupation with professional photography meant that he too needed a way of keeping his safari business going in his absence. By pooling our capital, we were able to afford a company manager who ran the day-to-day business of the firm. Eventually, in 1974, I withdrew from the company and its name was changed.

By the time of the wedding, I was becoming less enthusiastic about my planned entrance to university which would mean spending the greater part of the next three years in England. I had again become involved in field work and this provided further distraction. During my absence from Kenya, a young British geologist, John Martyn, had reported the occurrence of fossils in a variety of localities to the west of Kenya's Lake Baringo. My older brother Jonathan, who by now was married, was living at Baringo where he ran a snake farm. He had taken a natural interest in the fossil reports which he received and, on one of his visits to the geologist's camp, he noticed that amongst the finds was a fragment of human skull. Fossils of this nature are quite rare, so this discovery created considerable excitement and before long my father organized a small expedition to Baringo to search for additional fossils. While some of the team he sent worked at the site where the human skull fragment had been found, others searched the area for other sites and it was during this search that Edward Mbindiye found another human fossil – a jaw and parts of a skeleton, which to this day are considered to be of special interest. It was clear that the second site would require careful excavation and I was able to persuade father that I should take charge of this. I had just got married, I had no job and I needed to do something.

The second Baringo expedition was duly organized later in 1966 and I planned to conduct a full-scale operation at Edward's site. I was to be assisted by Margaret, who by then had a degree in archaeology and was so seen as capable of providing the appropriate academic guidance. The Baringo project was easy; the site was quite near Nairobi and there were no major organizational problems. I found myself closely involved with actual excavations and several happy months went by. During this period I also spent a lot of time exploring in the hills, looking for new sites and discoveries of my own.

One of the most impressive finds made at Baringo was that of a fossil elephant. This specimen was quite unusual because its skeleton was beautifully preserved. Usually carcasses were scattered, as they still are by scavengers, before they become buried and fossilized. The fossil was embedded in a very hard stratum of sediment that was tilted at an angle of some 30°, rather as if it had been put on show! It was so fine that I proposed that it should be excavated in a block and taken to the Museum in Nairobi for special exhibition. The task was sufficiently difficult and challenging to be especially attractive! Could I get back to Nairobi, undisturbed, a complete articulated skeleton? The whole specimen occupied an area about 14 feet × 8 feet and it was going to be a problem lifting this in a single block without damaging it.

Of course, on my own I would have failed, but fortunately I was not alone; I had the assistance of Paul Abell. Paul is an American whose profession is organic chemistry, but in 1966 he was on sabbatical leave and wanted to do something different. My father had proposed that he work with me on the Baringo Project and I quickly discovered his practical skills. In fact, Paul very soon took charge of the elephant and I found myself in the position of supervisor, showing up at the site to keep track of progress but in reality doing very little.

Our plan involved taking advantage of the tilted block which enabled us to excavate underneath the fossil. Trenches were cut round it and a timber frame was constructed and placed in the trenches. Working from the higher margin of the tilted block, the sediment below the skeleton was gradually removed and a timber platform was bolted plank by plank into position on the frame. To ensure the fossil was securely supported the small space between the planks and the fossil was packed tight with a mixture of sawdust and plaster of Paris. After about six weeks, our elephant was firmly established on its wooden platform, still tilted at 30° and ready to be transported to Nairobi.

This had been just the kind of challenge I enjoyed but while I had been so absorbed with my elephant the moment to apply for university admission had come and gone. I confess, I hardly gave it a thought! On our return to Nairobi I began to assist my father in his office. In 1961, father had resigned as Curator of the Nairobi Museum and had established an autonomous Centre for Prehistory and Palaeontology. He resigned from the Museum so that he could devote his whole time to prehistory and palaeontology and avoid the many administrative chores that go with the running of a public institution such as the Museum. The Centre was situated in buildings at the Museum and was affiliated to it. However, father had raised the money to run the Centre from overseas sources and he was, therefore, to a large extent independent. He was often away in America, usually for several months in each year, raising funds for one of his many projects and during his absence I would take over the administration. This kept me busy and I set to work to organize things as I wanted them. This was tactless and soon led to problems between us. Despite this, we retained a working relationship, and whenever my father was in Kenya, I would spend my time working on the collections we had made at Baringo.

It was during this time that I wrote my first scientific paper describing a new species of extinct monkey that we had recovered at Baringo. I was particularly fortunate to have as the subject of my paper a fossil skull and an almost complete skeleton of a very large and previously unknown form of monkey related to the colobus monkey. Because it was so complete and really quite unlike other fossil monkeys from eastern Africa, I was able to get away with a non-comparative and very basic piece of work. Nevertheless, I had to read a lot and in fact became very interested in fossil primates. I decided rather naively that this would be my area of specialization. My paper was duly published in a scientific journal and this gave me great satisfaction.

As a result of my association with the Centre for Prehistory and Palaeontology and the opportunities I had to be involved with its administration, I began to make useful contacts with Government officials, both civil servants and politicians. Consequently I learned that the Government was very unhappy with the way the National Museum was being run. This was partly the reason for the very poor level of financial support it received. At that time Kenya's National Museum consisted of just the Nairobi Museum and its very small staff. All the senior appointments were British nationals and virtually no plans had been made to train Kenyans to take over these positions. In fact, the total Museum staff in 1967 consisted of twenty-seven employees, of whom seven were at a senior level. My father's Centre was much the same, although he was Kenyan. Nonetheless the fact remained: no Africans were being trained nor were there any plans to rectify the situation. The Trustees were at that time far from convinced that a local Kenyan could measure up to the needs of the job and they were very reluctant to move towards the inevitable involvement of local people in the Museum.

All this provided me with a challenge I found irresistible; somehow I had to become involved. I decided to take advantage of the Museum's financial plight and I set about raising money for specific projects. To do this, I established an organization known as the Kenya Museum

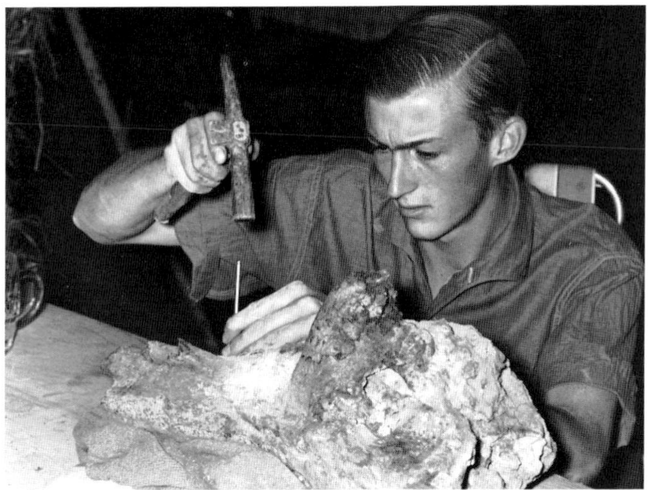

ABOVE Excavating the complete skull of a fossil pig at Peninj with Margaret Cropper, who was to become my wife

LEFT Removing rock in which the pig skull had been fossilized

BELOW The RAF Blackburn Beverley at Peninj after it had been flown in with my two vehicles in 1964

LEFT Discussing the Peninj australopithecine jaw with my parents and Dr Melvin M. Payne in 1965 at the offices of the National Geographic Society

BELOW A sequence of pictures showing the excavation and removal of a complete fossil elephant skull from a site at Lake Baringo in 1966

ABOVE Examining a crocodile, nearly 17 feet long, that we shot at the Omo. Left, Paul Abell; centre, Ndiki Kalulu; holding its snout, Allen O'Brien

BELOW Presenting funds to help build a secondary school at Kabete – my father's birthplace. The school Chairman, Mr Kiarie is receiving my cheque

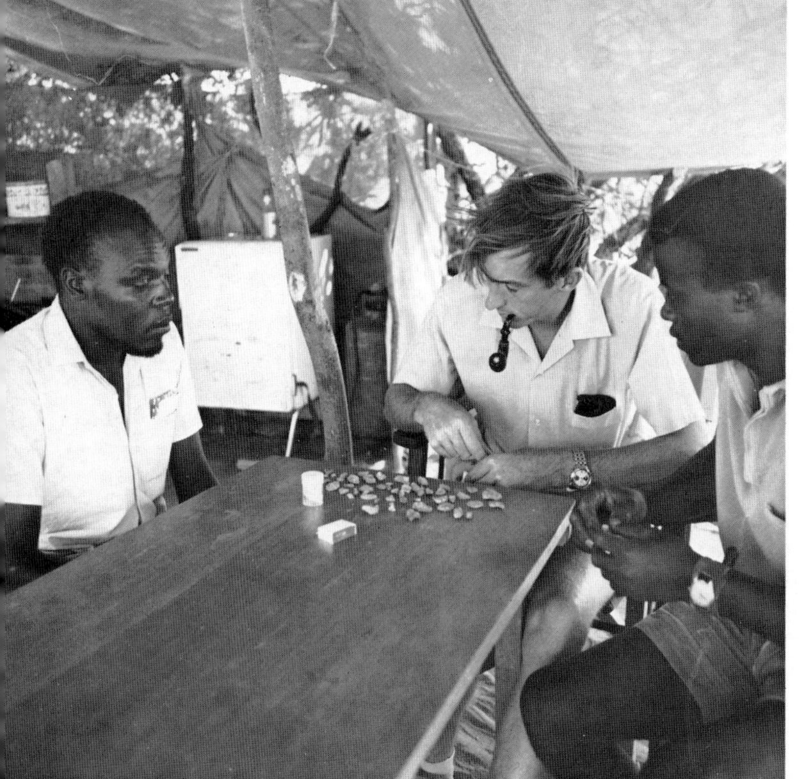

LEFT ABOVE Admiring the reconstructed skull and face of KNM-ER 1470

ABOVE With Meave, who is holding the skull of '1470' while I hold a complete femur that was found nearby in 1973

LEFT Examining fragments of '1470' with Kamoya and Bernard Ngeneo in our camp

Associates with its own board of directors and a formal constitution. I invited a number of prominent African personalities to become members, thus creating an important political contrast to the Museum Board which was then predominantly British and very colonial in outlook.

By virtue of my position as Executive Director of the Kenya Museum Associates, I found that I could attend the Museum's board meetings as an observer. At the time, the Associates were raising about ten per cent of the Museum's expenditure budget and so had to have a voice. I also managed to do some work in the display area by taking on the job of setting up my elephant exhibit. To do this, the prehistory gallery had to be completely reorganized and so I had ample opportunity to become fully involved in the institution's routine. By June 1967 I was very much a part of things and I had begun to make a nuisance of myself to its administration. I took great pains to keep in with Government officials and I expressed genuine concern that the National Museum was quite out of step with independent Kenya. I persuaded the Ford Foundation, with Government endorsement, to give funds for projects through the Associates, thus giving me the freedom to develop my own ideas for the Museum. I was able to recruit two young Kenyans, whom I then sent to the Smithsonian Institution in Washington, D.C. for special training as technicians. This greatly angered certain members of the staff who looked upon my efforts as gross impudence and interference.

My father, on the other hand, who had run the Museum for many years prior to 1961, seemed delighted with my efforts. He was, himself, not on the best terms with the Curator or the Board and anyway my activities were keeping me from interfering with his administration of the Centre for Prehistory and Palaeontology! My father was still very possessive about the place where he had done so much work between 1947 and 1961 and the Kenyan staff had remained very loyal to him. I was very aware that the Kenyans disliked the majority of the British staff at the Museum and I had the impression that it was not a question of skin colour but rather of national identity and it helped that I was seen as a Kenyan. (I had full Kenyan nationality at the time of independence and have proudly held a Kenyan passport ever since.) Here was a situation where Kenyan nationals were Kenyan with no distinction being made for skin colour. I felt very honoured and it made me all the keener to be truly Kenyan in outlook and loyalty. It was at about this time that I became politically aware and developed a strong antipathy towards colonialism and racialism.

One incident particularly incensed me and serves as a good example of the ridiculous attitudes that go along with racial prejudices. I was away somewhere but one of my trainee technicians was working in the Museum just prior to going off to America. He was a well-educated, mild-mannered, delightful young man who had just left a fine multi-racial school. He had not really worked at the Museum and had no reason to become involved with the senior European staff. During the course of the day, however, he needed to use the lavatory but to his amazement and embarrassment, and despite the fact that segregation had been abolished at the time of Kenya's Independence, he was warned by the Curator that this particular lavatory was not to be used by Africans. It was only available for the presumably 'clean' white bottoms of the senior staff! That this happened in Kenya late in 1967 is testimony to the strength of the colonial legacy that could so easily have permanently embittered the Africans. This, of course, was the sort of thing that provided me with ammunition in my fight with the Museum management.

The other important event in 1966 was that I bought a piece of land just outside Nairobi in the Karen suburbs and I began to build my own house. I supervised the construction myself to avoid employing a contractor and to save money. It was an interesting experience and it served to teach me a great deal about the building industry in Kenya; this has since been useful in my job at the Museum.

Again I found myself battling against the absurd legacy of British colonial rule. My architect had submitted building plans to the Nairobi City Council for planning approval, but I never really examined them in detail until I was well into the construction. When I did look at them I found that the plans called for air vents in each room; these I considered quite ugly and totally unnecessary – after all, what are windows for? Also, for no reason I could see, we were required to support the roof with an amazing quantity of timber, which would add considerably to the cost. I therefore took it upon myself to halve the amount of timber and omit the air vents.

The house was duly completed and the City Planning Team, without whose approval I could not get an occupation certificate, came to inspect it. After a very cursory inspection, a grim-faced bureaucrat informed me that the certificate would not be given, but that instead I would receive an order to pull my house down and start again. My house, I was told, had not been built according to specifications; the rooms were considered stuffy and so a health hazard, and my roof was unsafe. I managed to obtain a copy of the relevant by-laws and found that the roofing specifications stipulated that the roof should support three feet of standing snow, and the air vents were required so that the inhabitants would not suffocate if all the windows were shut to keep out the cold! The temperature in Nairobi seldom drops below 50°F; it has never snowed, and nor is it ever likely to. These were, in fact, English by-laws, totally inappropriate to Kenya, but nevertheless slavishly adhered to by officialdom.

I was certainly not going to pull down my house, and I sent a sharp letter to the City Council to tell them why. There was a long correspondence that became very threatening, but in the end, I was left to live happily ever after in my unsafe house. Although it was a storm in a tea cup, it does illustrate one of the greatest problems of the colonial legacy. When former colonies won their political independence from the European powers they inherited statute books full of out-of-date and quite irrelevant laws. It takes time to change laws and often the bureaucrats succeed in delaying desperately needed progress. Low-cost housing in the Third World cities is a case in point. Many excellent projects are rejected because they do not comply with out-dated regulations which, ironically, may have long since been abandoned in the European countries where they originated.

CHAPTER SEVEN

OUT OF MY FATHER'S SHADOW

DURING THE LATTER PART OF 1966, my father had been busy making plans for a new venture in southern Ethiopia, and my involvement with this provided plenty of distraction from the question of my own future. The plan was to mount a large international expedition to the Omo Valley. It was a site where work had already been done but much remained to do. My mother at this time was totally involved with work at Olduvai.

The Omo is a moderately large river that drains the Ethiopian highlands to the south of Addis Ababa and runs for some seven hundred miles before flowing into Lake Turkana in northern Kenya. For the last couple of hundred miles of its course the river flows in countless meanders with many small lakes cut off by oxbows. Here it is a moody, slow-flowing river, brown in colour from all the sediment that has been washed down from the highlands. The river banks are shrouded with luxuriant green forest while beyond lies the contrasting dry bush. It is part of an ancient drainage system – the rich fossil deposits in its valley are witness to this – but because the land is so flat, it is difficult to get an impression of the landscape unless you are in an aeroplane. Access to the Lower Omo Valley from the north is difficult at the best of times owing to very poor roads from Addis Ababa, but it is reasonably easy to reach the area from the south, travelling on Kenyan roads to the border and thereafter along tracks to the site. This route had been used before by scientists working in the valley but we required the approval of the Governments of Kenya and Ethiopia to use it again.

The Omo Valley had been visited by a French expedition in the early part of the century, led by a Comte de Boaz, who had collected a number of fossils which were eventually taken back to Paris for identification. As a result of the Comte's report, a second French initiative was undertaken in 1930 under the leadership of Professor Camille Arambourg, who collected more than a hundred tons of fossils over a period of two years! These were the subject of extensive study in the era before World War 2. During that war, my father took advantage of British troop movements in southern Ethiopia to arrange for his assistant, Heslon Mukiri, to spend a few weeks in the Omo Valley collecting for the Nairobi Museum. In the late fifties, Professor Clark Howell, from the U.S.A., arranged a short collecting expedition to follow up the earlier work but, unfortunately, his efforts were in vain because the Ethiopian border police confiscated all his fossils and made him return to Kenya. Clark, a close friend of my father's, was determined, somehow, to obtain Ethiopian approval for further work in the Omo. He was impressed by the immense potential of the area and the need for modern studies on the geology and palaeontology to supplement the earlier work done by the French.

It was not until 1966 that an opportunity arose. The Ethiopian leader, Emperor Haile Selassie, made a state visit to Kenya and our president, Mzee Jomo Kenyatta, arranged for my father to meet him. My father knew Jomo Kenyatta quite well although at times their relationship had been strained. During the difficult years of Kenya's struggle to attain freedom, the British authorities had arrested Kenyatta and charged him with a wide variety of offences related to the leadership and management of Mau Mau. My father, being Kikuyu himself, was called in as a court interpreter but before long he and the defence lawyer were at odds over the *real* meaning of certain phrases and words used in the proceedings. Before long my father resigned from the job and, in due course, Jomo Kenyatta was sent to prison. Quite naturally, father was rather concerned that he might be regarded with some disfavour but on the contrary, when Jomo Kenyatta was eventually released in 1961, their friendly relationship was soon re-established. At the meeting with the Emperor my father showed him various fossil discoveries from Olduvai and discussed the exciting potential for equally important sites elsewhere in East Africa. The Emperor was very enthusiastic and, because he was anxious that his country should also produce important finds, he invited my father to organize work in Ethiopia. It was agreed that any future expedition to the Omo should involve both Camille Arambourg and Clark Howell. From this the plan for an international project (Ethiopian-Kenyan-French-American) developed.

It was at this point in 1966 that I was invited to be a full participant. My role was essentially to serve as the field leader of the Kenyan team, and act as my father's representative. He was beginning to have a great deal of difficulty from his arthritic hips which made field work extremely difficult for him. My first suggestion was that we take a look at the site from the air. I wanted to see the extent of the site and to get an idea of some of the problems that might occur on the ground.

The valley has extensive outcrops of sediment which, it was agreed, should be divided into three principal zones. The first, allocated to the French, was the most southern and was the area where fossils had been previously collected by both Professor Arambourg and Clark Howell. The second area included deposits which seemed to extend north of the French zone to the river. This large zone, which had never been explored, was allocated to Clark Howell's team. The third zone was across the river. I was very happy to be allocated this area because, once again, it presented a tremendous organizational challenge. I had to get my vehicles and equipment across the water without using roads or bridges: a perfect African adventure! The first season was to be exploratory and future plans would be dependent upon what was found. It was understood by all that Clark's zone and mine might prove to be sterile of fossils in which case, in subsequent years, it would be necessary to co-operate with the French in their zone.

My father and I began to make plans for the Kenyan team and we decided to ask the National Geographic Society for support. Our request was met and during the early months of 1967 I had the task of assembling equipment and provisions for the expedition which was to leave Nairobi early in June of that year. I was given a great deal of help by an American friend, Allen O'Brien, who had been to Kenya on a safari while I was in England and, as a result, had become a close friend of my father and an admirer of his work. Allen was retired and keen to take part in the adventure; he was delighted to join my team and I gave him special responsibility for designing and building a ferry to take the vehicles across the river. In the few weeks before our departure, he busied himself in Nairobi constructing it and before we left, we successfully tested it on the Nairobi dam, where we could simulate the problems of muddy banks and deep water.

Finally everything was assembled and the expedition set off. The convoy consisted of three large trucks and some nine Land-Rovers. The French and American teams included various scientists and a number of Kenyans who were hired to help with camp duties. My team included Margaret, Paul Abell, Allen O'Brien, Alex Mackay (one of the Museum scientists), Bob Campbell for photography, Kamoya and a number of other Kenyans, some of whom were experienced in both fossil hunting and excavations. I was the undisputed leader of my own group, yet scientific matters were to be my father's prerogative. I had very clear instructions about maintaining regular contact with him back in Nairobi by radio and it was agreed that he would make several visits during our three months in the field. Also, Clark was near at hand and he would advise me if I needed scientific direction. It upset me to realize that I was not really independent because I lacked the academic qualifications and I began to have doubts again about my decision not to attend university.

On the third day after leaving Nairobi we arrived at the Ethiopian frontier from where we were accompanied by armed Ethiopian soldiers who had been provided by their Government to protect us. This part of Ethiopia was renowned for its armed tribes who had a reputation for discouraging visitors from penetrating their homeland. We soon arrived at the garrison town of Kalam, which also served as the administrative headquarters for the district, and was where the French planned to camp.

Ethiopians are to my mind amongst the world's most hospitable people and the official welcome that had been organized for us that day was a fine example. An ox and various goats had been killed, and an immense quantity of barely cooked meat awaited us. Bottles of whisky were provided to wash it all down. All this in one of the remotest parts of East Africa in the midday heat! Speeches of welcome were made and there was considerable gaiety before we finally persuaded our hosts that we simply had to move on. In the late afternoon our convoy struggled out of Kalam but inevitably after only a few miles we came to a halt. The huge meal and whisky had rendered a large part of our expedition unfit for the difficulties of working northwards through the trackless bush.

The next day we forced our way through the thorn bush for about fifteen miles before we eventually reached a point close to the Omo, where Clark established his camp. We went on in search of a stretch of river where the bank was not too steep and where it was wide and so comparatively slow flowing. We finally chose a crossing point close to a small village whose people became an interested audience. This was the first time that the villagers had been confronted with the trappings of an industrial society and our efforts gave considerable amusement. The large muddy river provided a home for an extraordinary number of crocodiles, many of which were huge, and we learned that they were largely mammal eating, taking significant numbers of cattle, sheep and goats as well as the occasional human who was foolish enough to venture near the water's edge.

The atmosphere of the Omo River that day was almost tangible. The river was sluggish, the air hot and humid, and thick gallery forest grew all along it; numerous birds were to be seen and in the great fig trees there were countless black-and-white colobus monkeys. We were a very long way from Nairobi and the next few months would require careful planning and a responsible leader. I was slightly daunted by the size and width of the river which I had previously seen only from the comfort of an aeroplane. If ever I had imagined myself romantically struggling to survive in tropical Africa, here was my chance to do it for real.

After several days of back-breaking work we had prepared a track down to the water's edge and built an exit ramp at a suitable spot on the opposite side of the river. Allen had put the ferry together and it was moored close to the ramp. The ferry's little nine horsepower engine had been tested in the Nairobi dam but not on a fast-flowing river, so we decided that it would be prudent to do a short trial without loading the ferry. It was lucky that we did because the current was far stronger than it looked and had it not been for a safety line the whole raft would have been lost! The plan to get the cars across had to be abandoned until we could obtain a more powerful motor, so I radioed to my father to send us one urgently. Until it arrived we explored the area across the river on foot.

I had brought along a small wooden dinghy and with this we moved as much equipment as we could across the river to the site that I had selected for our base camp. This was situated on the river bank underneath some magnificent African fig trees and overlooking the water where we would always see numerous crocodiles either swimming or sunning themselves. It was a very beautiful site on which to pitch our tents.

Initially, I was a little nervous about the crocodiles. For the first few trips up the river I kept the boat near the bank, working on the principle that safety lay in being able to get off the water as soon as possible if any crocodile took an undue interest in us. These river trips were quite unforgettable. From the boat we were able to watch the fantastic variety of bird and animal life along the forested banks. The colobus monkeys were everywhere and seemed quite unafraid; the crocodiles and hippos ignored us. Periodically we came across people who belonged to the small and isolated Mursi tribe who occupy this part of Ethiopia. The young women wore large lip plugs – clay discs of about three inches in diameter that were inserted in a slit in the lower lip. The discs were a form of adornment but they left terrible disfigurements in the older women, who commonly had great tears and scars where the plug had been. We found archaeological evidence which suggested that plugs of this type had been in use some 5,000 years ago. If this could be proved, it would be one of the few records of part of a people's culture coming through the Stone Age to the present in one area without change.

The men of the tribe seemed either to be off hunting with their ancient Italian Martini rifles or spear-fishing along the banks of the river. They stood immobile for great lengths of time with a long fishing spear held high,

A field camp on the banks of the Peninj River; dinner was cooked outside on an open fire

ABOVE Loading the raft at Lake Natron; my brother Jonathan and his former wife Mollie had been on a short visit

LEFT My parents and Glynn Isaac at the Peninj excavation, where the australopithecine mandible had been found

ABOVE At the northern end of Lake Natron, looking towards Ol Donyo Lengai. My brother Philip is on my right, in a check shirt, while Glynn Isaac is on my left

LEFT Surveying a camp site at Olduvai, Tanzania; the Ngorongoro Crater, south of Olduvai is in the background poised for action. At the slightest ripple they thrust the spear into the water. The success of their fishing was quite outstanding and from the archaeological records it was clear that harpoon fishing had been practised here for several thousand years. There was little evidence of game hunting but crocodile meat and hippo flesh were very popular. On one occasion a hippo had been killed in a fight with other hippos and a bloated and rather smelly carcass came floating down the river to become wedged on the far bank. The local Mursi people asked us to use our boat to pull the hippo across to a point close to their village. Bob Campbell did this and obtained some marvellous film of a hippo butchery.

We used the river a great deal, travelling up or down in our little dinghy to points from which we could explore inland. To begin with we only had the small engine, which gave us very little speed but it seemed to be adequate. However, after about a week we had an incident that made me change my mind. We had all grown accustomed to the hundreds of crocodiles and I, at least, had convinced myself that the noise of the motor was sufficient protection. I was not worried, therefore, about piloting the boat in mid-stream so avoiding the river's back currents and eddies.

One morning we left camp at about six o'clock, just before dawn. The colobus monkeys had begun their pre-dawn chorus and it was a beautiful time to be about. The light was changing rapidly and the pastel colours of the vegetation were screened by wisps of mist rising from the water. The first gentle rays of sunlight gave a golden hue to everything and, as anyone who has been to Africa will know, the early morning light defies description. As we slowly worked our way up-river we were mesmerized by the extraordinary beauty of our surroundings. I was steering the boat and accompanied by Margaret, Kamoya and Paul. Four people were the maximum our boat could manage and even this was overloading it. Suddenly the peace was shattered by an urgent warning from one of my companions and, looking back over my shoulder, I saw a large crocodile bearing down on us. There was no doubt, this monster was determined to attack presumably in the reasonable hopes of a meal. There must have been at least one hundred other crocodiles in sight and suddenly it seemed that all of them had developed a sinister interest in us.

Unfortunately we were mid-stream in a particularly wide stretch and it became very obvious that we would never make the bank in time. It also became instantly clear that if the reptile did get to us, it would make short work of our frail timber craft and our chances of survival would be very small indeed. What could we do? We were unarmed and had absolutely no means of defence. My first reaction was to attempt to turn the boat so that its higher prow would offer the most difficult target for our attacker. Unfortunately, the turn was too slow and almost immediately we heard the fateful splintering of wood as the giant reptile snapped at the side of the boat. I turned violently the other way, hoping to wrench ourselves free for a moment. This worked, but as my passengers were, quite naturally, now on the other side of the boat, we very nearly capsized. By some lucky chance we remained afloat; we had shipped a lot of water and we were now headed towards the bank. For a moment the crocodile seemed to have disappeared but to our horror it suddenly reappeared behind us, making for the stern where I sat. There was a tearing crash as it snapped its jaws shut. I looked round and there it was firmly attached to our splintered transom. To our great good fortune, the boat hit the bank at this critical time and, partly catapulted, we scrambled ashore, wet through and terrified. The crocodile angrily swam up and down, aware that it had been cheated of its breakfast but at last it disappeared under the water. Had it gone far, or was it waiting in ambush?

We soon appreciated our predicament. We were at least six miles from our camp and on the wrong side of the river – we just had to get back into the boat. Understandably there was little enthusiasm for this plan but as we talked we began to persuade ourselves that this crocodile was unique, possibly mentally deranged. Soon we had convinced ourselves that no ordinary corodile would attack a boat and thus fortified, and taking some branches as clubs, we re-embarked and set off for the opposite bank. Having crossed within a few moments without any sign of our attacker, we agreed to continue our trip up-river but this time I kept the boat near the bank so that we could escape should another attack occur.

In this manner we continued for about two miles and were just relaxing when another crocodile began a charge from about fifty yards away. This time we were on shore and well clear of the boat long before it reached us and although shaken we were in no danger. I now decided to abandon the river and return to camp on foot – there seemed to be rather more of these deranged crocodiles about than was healthy.

An attacking crocodile is a memorable sight when viewed from the safety of the river bank – from a small boat in the middle of a river it is unforgettable! On sighting a surface target the large crocs can swim towards their objective at an incredible speed. The only sign of their presence above water is a large head, followed by a huge bow wave like a torpedo's. The broad snout with a series of very large teeth shows quite clearly. Crocodiles hold their heads quite high, turning them from side to side the better to see their target. Like other reptiles their eyes are set in the side of their heads, which if kept still would not allow them to see directly ahead.

The march back to camp was not uneventful. First, we found ourselves in the middle of a baboon troop as we clambered up the steep river bank. These animals usually leave humans well alone but in this area they had no fear of us and for some moments I really thought we were going to be attacked. Then, having extricated ourselves, we set off along the footpaths and game trails and soon caught up with the fresh tracks of a pride of lions going in the same direction. As you might imagine we slowed down and were greatly relieved when the lion tracks branched off, leaving our own way clear. It was late that day before we got back to the camp.

By this time the larger engine for Allen's pontoon had arrived and we had successfully transported the cars across the river. So the next day we took the pontoon up-river to bring back our little boat. Although very slow it was quite safe and the mission was accomplished without incident. After the crocodile attack I sent an urgent request to my father for a slightly larger aluminium boat. With this, and the new forty horsepower engine that we had for the ferry, we were again able to use the river. As we had discovered, crocodiles over sixteen feet long would have no fear of attacking small boats but from our new boat we could taunt and tease them in safety to our hearts' content.

In the area we were exploring there are extensive outcrops of sediment which from the air had looked like the kind of stratified silts and clays where we might expect to find fossils. On the ground, however, we soon found that most of these were of quite recent geological age, spanning the past 150,000 years or so but no more. Our hopes had been to find much older strata where important palaeontological finds could be made. There were some much older sites, perhaps as much as 3.5 million years but these yielded little of interest. Even so, we did make some exciting discoveries and we managed to collect a number of interesting fossils and stone implements from the later period.

Quite early in the expedition, Kamoya found several pieces of a fossilized human skull and further excavations yielded parts of a skeleton. A few weeks later, Paul Abell found a second fossil skull in the same geological strata on the other side of the river. These two skulls have since proved to be quite important as they represent a stage in human evolution that is still not particularly well documented.

Geological investigations and dating have shown that the two skulls are about 130,000 years old, yet despite their antiquity they are both clearly identifiable as *Homo sapiens*, our own species. At the time of their discovery, scientists generally believed that our species had only emerged in the last 60,000 years and many considered the famous Neanderthal Man to be the immediate precursor to ourselves. The Omo fossils thus provided important evidence that this was not so. Although I had hoped for much older material, I was well pleased. My parents flew up to see our work, and they too felt our expedition had been worthwhile. However, we could not really justify a further season the following year in the 'Kenya zone' of the Omo Valley. Meanwhile the French team, under Professor Arambourg, was reporting great success with lots of fossils, among which was an australopithecine jaw, older than the material then known from Olduvai. Clark Howell's party had also found some rich fossil beds of Pliocene age between 3.5 and 2 million years old, as well as extensive deposits of the later period.

The three teams had contact with one another by radio and every so often a meeting would be held to discuss progress and consider matters of mutual concern. Although it was a joint international expedition each team had a clear sense of its national identity. I always felt there was a slight undercurrent of competition, though perhaps this was only me being unduly sensitive. I was conscious that my team was doing less well than the others. Once again I found myself frustrated. My career seemed to be making no real progress.

Periodically we chartered a small plane to bring up supplies. On one of these flights I took the opportunity to return to Nairobi for a week or so to attend to matters in my safari company and in the Museum. It was on the return flight from Nairobi that I first saw the site that has occupied my time ever since. I was a passenger on the small six-seater plane and as we approached from the south what was then called Lake Rudolf and is now Lake Turkana a huge thunderstorm was in progress over the western shore of the lake, directly in our flight path. The pilot decided to detour to the east and this took us up to and along the eastern shore which I had never seen before. Fortunately we were fairly low and I was able to see clearly large areas of well-stratified sediments. I was quite surprised because the maps of the region showed nothing but volcanic rock for the whole area. I was wildly excited by my observations and determined to visit the area as soon as possible.

Upon my return to my camp at the Omo, I made contact with Clark Howell. The American party had made arrangements to charter a small two-seater helicopter to move people from site to site and so avoid the terrible conditions of almost impenetrable thorn bush that they would otherwise have to drive through. I asked Clark whether I might hire the helicopter for a day to do some prospecting. He agreed and a date was arranged. On the appointed day, the helicopter collected me at seven o'clock in the morning and we were quickly on our way south after crossing the Omo river and into the dry country to the northeast of Lake Turkana that I have since come to love so much. Before long we spotted some promising-looking sedimentary outcrops and put down to investigate. I jumped out of the helicopter and immediately found some fossils as well as stone artefacts which had presumably eroded from the strata. We explored further, landing at various other outcrops and everywhere we stopped there were fossils. I was absolutely delighted. My mind was made up; I would return to this Lake Turkana site the following year with my own expedition.

Back at the camp, I excitedly shared my news with the Kenya team, all of whom were eager to know where I had been. To my horror, I discovered that in my haste and excitement I had completely forgotten to locate the fossil sites on any map and I had only a general idea that we had been somewhere northeast of the lake. Were we in Ethiopia or Kenya? I had no idea but felt sure that the deposits would extend into Kenya for some considerable distance.

By late August 1967, we had completed our survey north of the Omo river and we and the other teams returned to Nairobi. The recrossing of the Omo proved relatively easy and the journey home was uneventful. We took with us the two skulls we had found as well as animals' teeth, horn cores, etc, for study at the National Museum's Centre for Prehistory and Palaeontology. All the material was of Pleistocene age, dating, in other words, to the period when our immediate ancestors were evolving. As before, with the results from both Natron and Baringo, I still lacked the credentials to write the scientific report and I had to invite others to assist. The French and American teams made their own arrangements for a long-term programme of research at the Omo while I began to anticipate and plan for my own expedition in Kenya in 1968. So ended my formal association with the Omo Expedition and my work in Ethiopia.

In January 1968, I again went to the United States and visited the National Geographic Society in their Washington headquarters. My role was to give a report to the Committee for Research and Exploration on the Omo Expedition and, in particular, to detail plans for further work there. Father was in Washington too, reporting on the progress of work at Olduvai and other sites. He expected to submit a formal request for

$25,000 to enable me to return to the Omo in 1968. Although I had discussed the idea of my new project with him, he was not especially enthusiastic and wanted me to continue with the Omo. In part, this may have been because of his personal commitment to Emperor Haile Selassie, but I also think he felt strongly that I would benefit from working under Clark Howell and Camille Arambourg. Of course, this was one of the main reasons why I wanted my own project; I was determined to put an end to my working 'under' people. At 23 I was an impatient young man without any time for advice.

The Committee for Research and Exploration of the National Geographic Society holds its meetings in a magnificent boardroom. It was here that I solemnly presented my report on the Omo after my father had spoken about Olduvai and his view of the long-term potential at the Omo. I introduced the idea of my new project at the end of my report, stating my wish to investigate the eastern shore of Lake Turkana and specifically asked the Committee for permission to use the requested $25,000 for this, rather than returning to the Omo as had been suggested. There was I think, some surprise, and my father certainly spoke up and said that he would favour the Omo plan where at least we knew fossils were to be found in quantity. I think he also referred to the younger generation having no patience to continue a project that had not produced immediate prizes. He was right, of course. The committee formally excused us after our presentations, and the Chairman, Dr Leonard Carmichael, invited us to join the Board for luncheon later in the day. It was during lunch that I was told that my request had been approved, but I was also given to understand that my reputation was on the line. I had argued for the opening up of an unknown area and had resigned from the established Omo Project. I was delighted: at last I was allowed to initiate a project independently of my parents. For my father, though, it must have seemed very different because money that would otherwise have been used to continue a project which he had begun and in which he would have continued to be directly involved was to fund research in an area that at the time was completely unknown. For me it was a gamble and I had no idea of how things would work out. I now know that I made the right decision but for quite the wrong reasons.

By the end of May 1968, I had thoroughly prepared my expedition to the northeast of Lake Turkana. I had again flown over the area, this time to plan our route and I had invited certain young scientists, including Bernard Wood, John Harris and Paul Abell, to join me. My wife, Margaret, who had been involved with my plans for Lake Turkana from the very beginning was also to be a member of this new expedition.

I was elated at the prospects of a three-month adventure to the spectacular and desolate land of northern Kenya. I had my own funds from the National Geographic Society, and the expedition was mine. The success or failure of the project was going to be laid on me and I was delighted with the responsibility. I was sure that we would have a good time and my original helicopter visit had, as far as I was concerned, established that we had a very good chance of finding something exciting.

The northern parts of Kenya are characterized by a semi-desert environment with spectacular vistas. It is a harsh land, particularly for those who are intimidated by the vastness of Africa. For me, the deserts of northern Kenya provide some of the finest expressions of the continent's charm. The constantly changing colours of the arid land, the twisted sculptures of dying trees and ancient lava flows all combine to create a picture which is impossible to describe adequately. The fierce winds that are generated from the conflict between the searing heat, beating off the blackened volcanic rocks, and the cool air from the water of the lake make the whole place seem alive and restless. These winds reach almost hurricane speeds as they cross Lake Turkana. The simple, sharp contours of numerous volcanic cones are reminders of the turbulent recent history of this area, and one wonders if, some time in the future, there will be further outpourings.

Across this area wander the Gabbra, a nomadic people who herd their camels and goats between water holes and sparse pasture. They are a proud, fierce people typical of desert dwellers the world over. Their survival has been made possible by their extraordinary discipline and stoic adaptation to one of our planet's harshest habitats. There have been few incursions by our technological society in the African deserts except where precious minerals or oil have been found. The scarcity of water and the almost complete lack of rain from one month to the next almost guarantee that the deserts will be the last unspoilt wildernesses of our planet.

It was indeed a privilege to plan an expedition into this exciting area. I was eager to be off but there were many preparations and one of these was to assure the safety of the expedition from attack by the fierce raiding bands that had gained such a reputation in the area near the Kenya-Ethiopia frontier. I made some tentative enquiries of the police, who were quick to inform me that I would not be allowed to proceed into the area at all. This was an unexpected blow and I set to work to find a way round it.

I arranged for Kenya's Minister of State in the President's Office to come to the Museum on a formal visit. This Minister had considerable influence in the Government and was a close friend of my father; the two men had known each other since childhood and belonged to the same age-group in the Kikuyu tribe. According to Kikuyu custom, I enjoyed certain rights of access, not unlike the Christian system of Godfathers. Whilst the Minister was visiting the laboratory where the technicians were making casts, I took the opportunity to point out to him that important work was in progress in Ethiopia and Tanzania but that no significant sites were known in Kenya. This appeal to his national pride had the desired result. I explained that approval for my plans would put Kenya firmly on the map in terms of studies of early human development. Apparently I was convincing, for the Minister promised to help and after a few days I was informed that the Government would provide an armed detachment of police to escort the expedition for a period of three months.

I had no idea of the real conditions at Lake Turkana, nor in fact did I know exactly where I was headed. I had a notion that I should use the three months to explore fully the area to the east of the lake, beginning about half-way up the shore at a feature known as Allia Bay. From an aerial survey that I had made after the helicopter trip in 1967, this seemed to be the ideal place from which to start our operations. I planned to have a small boat for use on the lake and Allia Bay offered the prospects of some reasonably sheltered water. My idea was to take the boat up the coast and make a series of easterly traverses from the shore. I realized that overland travel by vehicle was likely to prove difficult, especially in the initial exploratory phase, and there was also the problem of restricted fuel supplies. Although I had previously organized many safaris, the trip to Lake Turkana was to be my first experience in the semi-desert.

The convoy to Turkana in 1968 included a large four-wheel-drive truck, two Land-Rovers, and a Ford Bronco. We towed the small boat on a trailer. The journey north took us through Kenya's Rift Valley to Nakuru from where we climbed out on to the shoulder of the Rift and then on to Maralal, a dusty village on the edge of the northern district. At the end of the second day we picked up our police escort at the village of Baragoi and at about noon on the third day we came over a barren lava ridge to look down for the first time on the glistening water of Lake Turkana which, because of its colour, is often known as the Jade Sea. We began to feel closer to our destination but from this point on our progress became slower and slower as we coaxed the vehicles over the rugged landscape. Eventually we arrived at the southern shore at a place called Loyengalani where we stopped for the night. We had covered less than a hundred miles that day.

ABOVE Collecting fossils with Meave while Ngeneo, Masau and Nzobe discuss strategy for continuing the search

LEFT On the verandah at Koobi Fora with Kamoya. The various modern skulls in the foreground are used for comparing with fossils at the site

BELOW It was always necessary to refuel the Cessna before the return flight to Nairobi

ABOVE An aerial view of the landscape at Koobi Fora

BELOW Going somewhere with Kamoya – probably to see a new fossil find

LEFT Leading a Wildlife Club's demonstration in Nairobi

From Loyengalani the track north became almost impossible to follow and progress was extremely slow. I knew that during World War 2, British troops had moved northwards to the frontier of what was then Abyssinia and I hoped to make use of these roads. Unfortunately, more than twenty years of disuse had meant that many sections were completely obliterated and we had to find our own way, hoping eventually to rejoin the old track somewhere ahead. In the arid desert conditions, however, gun emplacements and stone walls for bivouacs had been well preserved. Our advance was very laborious and on the fourth and fifth days we covered just forty or so miles in nine or ten hours of effort. Eventually, on the sixth day, late in the afternoon, we reached Allia Bay where we stopped by the shore, thankful to have arrived at last. After a hurried meal, everyone had an early night, simply setting up beds under the brilliant star-filled sky. During the night the wind came up and by early morning it was blowing at thirty or forty miles an hour. Thank goodness we had been too tired to put up tents the evening before; they would surely have been blown away during the night.

I was impatient to begin our exploration, but before we could do so we had to spend several days establishing camp, launching the boat, making an airstrip, and locating a source of fresh water. The lake is rather alkaline and, although the water can be drunk, it has a strong laxative effect and it produces undrinkable tea. Fortunately, Kamoya located a deep lava pool at Karsa, not far from Allia Bay, and although the water was very green from the algae it was sweet and in plentiful supply. We subsequently discovered from the carefully laid lines of stones that this had been the site of an army camp during the war. It was also a traditional water hole for the nomads who used the area from time to time.

The airstrip was easy to make and we marked out a level piece of ground which still serves today as the main airfield for the area. The tents were another problem. We had to build stout wind screens by binding bundles of marsh grass on to wooden frames, which provided enough shelter to enable us to keep up the eight or nine tents. Nevertheless, there were tremendous wind eddies every night, which filled each tent with dust and sand.

By 10 June we began, at last, the systematic exploration of the area. As I had expected from the evidence I had found earlier when I landed in the helicopter, there were fossils everywhere and from the first day we began to find a marvellous selection of animal remains. I could not have asked for more; my optimism had been completely vindicated. The terrain was very rough indeed and we used the vehicles sparingly, preferring instead to walk. On some days we must have covered twenty or so miles although usually it was much less. The boat that we had with us also proved very useful: a number of sites where we have since found important specimens were first reached in 1968 by walking inland from the coast.

As we explored this incredible area, it became increasingly obvious that we were quite alone. For day after day we never saw any indication of human life at all. The police told us that the area was empty because the local people were afraid of the raiding parties which came down from the north in search of cattle and sheep.

In contrast to the scarcity of humans, there were plenty of animals. The grassy zone along the lake shore and the well-vegetated banks of the sand rivers supported a wide variety of game, including zebra, topi, giraffe, oryx, ostrich and Grant's gazelle as well as various predators such as lion, leopard, cheetah, wild dogs and hyenas. It was great fun to be on foot amongst the animals and this heightened the sense of adventure which I so enjoyed. Every day was filled with incidents, many of which were unimportant in themselves but which were like delicate spices on a good meal, adding tremendously to our sense of being involved in a very special experience.

Each day we got up well before dawn. I would normally wake at five o'clock and make the rounds of the tents, calling the others to rise and come for breakfast. Usually by six, everyone had downed their tea or coffee and some bread and we would be moving off, either in the boat or a Land-Rover, to explore a new locality that we had chosen the previous day. This routine was followed for six days a week, but the seventh was always a day of rest. I continue to operate this system simply because I appreciate the value of a change in routine and not because I attach any significance to the Sabbath. Indeed, the day off may not necessarily be on a weekend at all but a complete rest or change of pace is very important.

By the time the sun was up and the light was good enough to see clearly, we would have arrived at the point from which we could begin our prospecting. In this way we had the full advantage of cool mornings and by noon we had usually completed at least five hours of useful work. The period between noon and three o'clock in the afternoon was much too hot to search for fossils and so we would either return to camp or rest in whatever shade could be found. In the late afternoons we would continue our exploration.

One of our biggest problems was the sheer quantity of fossils! The specimens were so numerous, many beautifully preserved and complete, that I realized we could never collect them all in the first season. Consequently I decided to collect only a few examples of each species as a basis for convincing our sponsors of the need for further expeditions and I also wanted some 'trophies' to show Clark Howell who was still working at the Omo sites. Our principal objective was, of course, to find remains of early man. In the late sixties there were not very many east African discoveries and very little was, in fact, known about our evolutionary record. Olduvai and Lake Natron were the principal sites and the former had been searched for thirty years before the first *Australopithecus* was found. I had been extremely lucky at Natron and I was determined to prove that luck was still with me. Kamoya and three other Kenyans were the core of my prospecting team and they were well aware of the importance of a hominid find to our future. Fortunately our first hominid fossil was discovered in the initial three weeks; it was a very poor specimen of a massive lower jaw of *Australopithecus boisei*. Even though it was not in good condition – all the teeth had been lost and the bone was very weathered – we had proof that hominids were indeed to be found in the area. With such a discovery we could relax the pace slightly and concentrate on a careful survey of the fantastic area that we had come upon.

After about three weeks we had covered the area that could easily be reached from our base camp and it made better sense to move camp. I was particularly keen to find out the full extent of the area and so I decided to go northwards and to continue our exploration until we either reached the limit of the deposits or the Kenya-Ethiopia frontier. I had no permission to work over the border, although I am sure that it could have been arranged. Camp was moved to a water hole called Nderati some fifteen miles to the north. It was perfect for our requirements, having ample water and plenty of shade. There was a series of fairly shallow man-made wells surrounded by a forest of desert palms known as the *Doum* palm. These wells were and still are used by nomads from time to time. There was again evidence of wartime occupation around the spring. A small fortress-like ruin sits on one of the hills near the oasis and in the surrounding area there are numerous signs of old roads and look-out positions. The palms provide a brilliant green patch in an otherwise stark, almost purple landscape of volcanic hills and boulder-strewn plains. Vegetation is scarce, and only comes to life in a burst of almost springlike activity after one of the rare rainstorms in the area. The contrast with our lake-shore camp could not have been more dramatic.

We nestled our tents in amongst the various trees. We had so much water that we set up canvas showers for bathing at the end of the hot and dusty days. It was such a luxury to come from a very cold shower and sit under the brilliant stars, talking in hushed voices about the day's events. I don't think I have ever enjoyed the hour before dinner quite so much as on those evenings at Nderati.

The camp was well guarded against bandits. Our police escort set up positions on the small hills around and we were kept under constant watch. At night, sentries patrolled and it was necessary to stay in the tents or carry an obvious light for immediate identification. During the day, a detail of about seven men would be attached to the prospecting team and as we explored, our armed guards would keep watch, moving from hill top to hill top. It was a strange experience and at times nerve wracking.

On one occasion, as we were driving back to camp in the heat of the day, someone spotted what looked like a group of men crouching under a tree some two hundred yards away from the track. The heat haze was so strong that it was impossible to see clearly but we decided to investigate. We drove to a position out of sight and I set off with the policemen, intending to encircle and capture the 'bandits'. I assumed that the people we had seen were bandits (a term I use rather loosely) because we were in the unoccupied frontier zone – an area without villages – and so anybody seen there was likely to be engaged on a mission of some sort. Certainly, the few people we had previously come across had been armed. My police companions were very reluctant to accompany me, preferring first to go back to camp for reinforcements. After some thirty minutes of painful stalking, I was in position to lead a charge over the ridge and thus I hoped to surprise the group of men whom we were sure were still under the tree. With a shout of 'Forward' I heroically led the charge only to find that our bandits had turned to lava boulders! There were about seven large rocks under the tree and their shape in the heat haze had completely fooled us. We felt very silly but thankful that we had not gone back for a machine gun and further troops!

By the end of August we had explored an enormous area extending from Allia Bay northwards to the Ethiopian border, and from the lake shore inland some twenty miles. It was a vast fossil site, covering about 500 square miles and at that time it was certainly the largest site of Plio-Pleistocene age that had been discovered in Africa. The abundance of fossils was extraordinary and I realized then that we had stumbled on to something far bigger and far more important than anyone suspected. By the end of that first season we had found two more hominid specimens, giving a total of three, which far exceeded our most optimistic expectations. The only disappointment was that I had completely failed to relocate the site that I had first visited by helicopter – the real goal of the expedition – and also we had not found any stone tools.

When we got back to Nairobi and unpacked the fossils, I had the fun of seeing people's reactions. My parents were delighted and indeed father congratulated me on my decision to go to the east side of Lake Turkana. He told me that he had always wanted to explore the area but lack of funds and equipment in the early years had made it impossible. I was never quite sure whether this was really true because he certainly never mentioned it before the expedition; in any event I decided not to believe it because I wanted to feel that it was entirely my site. It was then extremely important for me to prove myself independent from my parents.

My colleagues on the Omo expedition returned to Nairobi in September, at the end of their second season, and they too were very complimentary at our success. Indeed Clark Howell urged me to go to university and to write up the fossils we had found as part of a doctoral thesis. Clark undertook to help get me into an American university and offered to serve as my supervisor. He also offered me the opportunity to write up the fossil monkeys from his Omo collections because of my interest in non-human primates. At that time I had decided not to involve myself with the hominids, partly because I thought they were too important for someone who had no proper credentials, and also because my father was a world authority on early man and I did not wsh to come into conflict with him. I accepted Clark's offer of the fossil monkeys but declined his suggestion that I register for a university degree. I saw no reason why I should not continue to work my site as I was and so I made the final decision to make my mark as a scientist without undergoing a formal university training. I have never regretted this choice. Clark later offered the fossil primates to another young man who, under Clark's supervision, wrote them up for his doctoral thesis. I gradually became more involved with the fossil hominids that we had found and the monkeys from my site east of Lake Turkana were studied separately from those at the Omo.

And so it was that I began a project that has, as I hoped it would, put Kenya firmly on the map in terms of Early Man studies. Each year since 1968 there has been scientific work at the site, and today a permanent research establishment is based there.

CHAPTER EIGHT

IN CONFLICT AGAIN

EARLY IN 1968, I began a campaign to have myself given an official position in the Museum's administration. I had returned from America and I was planning the trip to Lake Turkana in June. I had realized, however, that I needed a job with a steady income because I could not spend all of my time in the bush looking for fossils. In 1966 and 1967 I had been helping my father with his work in the Centre for Prehistory and Palaeontology, where I had been an administrative assistant and had become interested in looking at bones from a scientific point of view. I was anxious to see changes in the Museum, too, and quite sure that only I could do what had to be done. As is to be expected, the Museum Board and its Director were appalled that a twenty-three-year-old, untrained, high school drop-out was campaigning to take over the National Museum! Finally, Joel Ojal, then the Permanent Secretary in the Government Ministry controlling the Museum, gave an ultimatum to the Board Chairman. Either the Museum took on Richard Leakey and made efforts to Kenyanize senior posts or else all government funding would cease. Not surprisingly, under these circumstances, my chances of a job were greatly improved. It helped that Joel was also Chairman of the Kenya Museum Associates and I was thrilled that I was being supported by the Kenya Government as a Kenyan to replace a foreigner! I had achieved one of my ambitions; to become identified as a loyal national of my country regardless of my European origins.

After several months of painful consideration, the Museum Board offered me the position of Administrative Officer. I was told that there was not enough work for a full-time administrator but that I would help the Director on a part-time basis as well as continuing to help my father in his Centre. I was also given to understand that I should have to grow a lot older before I could be seriously considered for a senior position. My complete lack of academic training and qualifications were used as a strong reason why I should not take on any responsible role. I was indignant; I felt that I was being unfairly discriminated against because of my age and so I made up my mind to fight a final round. By 1968 the Museum Board included a number of Kenyan members who had been appointed by the Minister upon my recommendation. It was my hope that the matter would be dealt with as a political issue and I argued that the British Chairman was deliberately frustrating reasonable efforts to Kenyanize the National Museum administration.

I duly attended a Board meeting in May 1968 at which I was expected to accept the offer of a position as Administrative Officer. The meeting assembled and after a few preliminaries I was called upon to give my response. I calmly stated that I could not accept the post as offered and I urged the Board to upgrade the position to Administrative Director with full responsibility to the Board for all the administration of the National Museum. The Director threatened to resign if this was done and several Board members gave warning that my appointment would be a disaster. I could see that a number of Kenyans were ready to fight my cause and so I was relieved when the Chairman excused me from further attendance at the meeting.

The Board's decision went my way. They offered me the position of Administrative Director while accepting the intention of the incumbent Director to leave in protest at the conclusion of his contract the following year. I was urged to use the period before his departure to learn the job as well as possible and I was given clear directives to keep away from the scientific aspects of the Museum's work. I was reminded that I was not a scientist, that I had no credentials and that I was unknown. The Board planned to appoint an experienced and prominent person to take charge of the scientific aspects of its responsibilities. This was no threat to me because I knew that whoever was in charge of the finances was in fact in control of everything and once the Director had left I would be completely in charge. I also saw no reason why my job at the Museum should not provide me with a platform from which I could gain academic recognition in due time.

I agreed to accept the post, effective from the beginning of October 1968 when I expected to have returned from the eastern side of Lake Turkana. To this day I do not know what was said at that fateful Board meeting except what is in the very formal minutes, but I do know that the Kenyans stuck together and gave me the necessary support. My father was a member of the Board at the time and I am proud to know that he also gave me his full weight, even though he must have realized that I was taking up a position which could create difficulties between us, especially over the independent Centre that he ran.

This, in fact, happened sooner than I anticipated. I became involved with negotiations between the Centre for Prehistory and Palaeontology and the Tanzania Government for the return of some of the Tanzanian collections. Up until that time, all the fossils and stone artefacts collected from Olduvai and Peninj had been stored in Nairobi and this was justified on the grounds that they were being actively studied. My impression was that, while some parts of the collection were indeed being worked on, a large quantity of material was simply taking up precious storage space. My father's view was that it would be preferable to keep all the Olduvai collections in one place and the best solution was to maintain the *status quo*. I actively disagreed with this and felt strongly that there were excellent political as well as practical reasons to return the material that was not being studied to the National Museum in Tanzania.

For some reason, father was unable to attend a crucial meeting in Tanzania on this matter and, despite the fact that my views differed from his, he sent me as his official representative with full authority to take decisions. I think he believed that I would represent his views faithfully and negotiate a reasonable arrangement. I am afraid that I quickly agreed with the Tanzanian position and formally undertook to return some thirty crates of material. This decision led to a tremendous row with my father and I am sure that had I not been his son I would have lost my job in the Centre – I nearly did, anyway. I felt then, and still do, that archaeological and other scientific collections must always remain the property of the country of origin and that provided safe and effective storage can be arranged, collections should reside where they belong. This concept is especially important in developing nations, because it contributes significantly to the build-up of a scientific identity. Scholars from various parts of the world have to visit these institutions if they are to study collections and, in this way, the local scientists gain useful contacts as well as being able to have some control on the research carried out. The importance of this is seldom understood by Western scientists who tend to see such ideas as an attack on the internationalism of science.

One important aspect of archaeological and palaeontological work is that it is possible to make very accurate copies of important specimens as casts. These are generally used for museum exhibits, teaching and research. The casts were made in plaster of Paris, but now materials such as fibreglass and plastics are used. During 1968 when my father was away from Kenya I took advantage of the presence of a British lab-technician who was employed to prepare fossils and to make plaster casts of material at the Centre. I had learned that my father planned to send some important material to Britain for reproductions to be made, claiming that there existed in London a very sophisticated firm which could cast in plastic. Father was quite adamant that these skilfully made casts could not be

Waiting for the raft to arrive,
Lake Natron

With a Msonjo guide in long grass, accompanied by Margaret and my brother Philip. Because of rhinos I always carried a gun

My brother Jonathan at the Wasonjo village of Sale, west of Lake Natron

Dawn near my camp on the shore
of Lake Baringo, 1966

produced in Kenya and it was for this reason that he had entered into an arrangement with the British company. I was unhappy with the idea that a British firm was to make a profit on Kenyan fossils and I had worries about the security of our specimens travelling to and from Kenya. I was also unreasonably angry when I learned that the firm, F.D. Castings Ltd, was in fact solely run and owned by Jane Goodall's sister and that the initials stood for Foster Daughter. Jane Goodall was a protégée of my father's. Over a period of time, my father encouraged three different women to study the behaviour of wild apes. Jane was the first and she became famous for her work with chimpanzees. Later Dian Fossey worked with gorillas and Birtué Galdikas-Brindamour studied orangs in Indonesia. Jane's family more or less adopted my father and he often stayed with them when he was in London. They were very fond of him, and he of them, and he helped Jane's sister Judy to establish her casting firm which, in the beginning, operated from a room in the Goodalls' London home. Looking back on it now, I cannot entirely justify my personal resentment but at that time I was a very jealous young man. I certainly felt that my father should have been helping me to set up a casting programme in Nairobi to benefit Kenya rather than taking the work to Britain, but then he was a Kenyan with dual allegiance whereas I was very nationalistic.

Anyway, while my father was away, I asked the technician whether he could make plastic casts in Nairobi, and to my delight he said yes. I asked him to make some examples and to train some Kenyan technicians. By the time father returned to Nairobi, I had in hand some excellent coloured plastic casts and a strong case for not employing F.D. Castings. Again I was in direct conflict with my long-suffering father. He was furious for a while and probably deeply hurt, but that made no difference to me then. As it happens, the sale and distribution of casts from the Museum has since earned it an excellent international reputation and quite a reasonable income. Once more, I was doing the right thing but largely for the wrong reasons; in other words, I was acting at least partly out of spite even though the programme I initiated did benefit the Museum and Kenya.

In October 1968 I started my job as Administrative Director of Kenya's National Museum. I realized that I had a wonderful opportunity to establish my own reputation and status. At that time, the Museum was small and poorly funded and there seemed to be tremendous opportunities to create an active and useful national institution. The opposition to my appointment by senior and well-established staff at the Museum, and the board's reluctance in employing me, left no doubts in my mind as to the difficulties ahead. Indeed, when I asked for an office and a secretary, I was told that such trimmings were quite unnecessary and impossible to arrange. I was to work where I could, and any letters that I had to send were to be typed by the Director's secretary who had also been opposed to my appointment.

To get around all this, I persuaded my father to let me have the use of a room in the Centre for Prehistory and Palaeontology which was adjacent to the Museum. I persuaded a very attractive young American lady to be my unpaid secretary who, though not a very good typist, certainly gave me status! Everything was arranged and I was due to begin the job on 1 October. Three days before this I became unwell. My illness did not seem especially significant on the first day when I merely developed an extremely sore throat and a fever. On the second day, I awoke with a very puffy face and obvious signs of fluid retention. The doctor I went to see diagnosed kidney disease, stimulated by the throat infection, and he recommended that I take a complete rest for up to six weeks, explaining that this would be the only way that I could hope to avoid serious damage and ultimately fatal destruction of the kidney tissue. I was numbed by the news but I did not really believe the doctor. Also I had to consider the fact that if I failed to turn up for work on my first day and then was absent for six weeks, the Museum authorities would certainly withdraw their offer of my appointment. The job was essential to me and crucial to my ambitions and I could not possibly consider the prospect of losing it. Consequently, I spent only a morning in bed and thereafter went on with my normal activities. Apart from a dull pain in my back, the obvious signs of kidney problems soon disappeared and I was at my desk as planned on 1 October.

I decided not to tell anybody about the problem except for Margaret who was appalled at my decision to ignore the doctor's advice. She did not believe I should have been offered or have taken the job at the Museum in the first place, feeling, I think, that I was far too young and inexperienced to take on such a potentially important post. She had a point, of course, but her attitude upset me. In March 1969 I passed through London on my way to America to raise support for the Lake Turkana expedition and I went to see a specialist about my kidneys. I was told that they were permanently damaged and that I could expect them to deteriorate. This time I believed the doctor and I was a little frightened. I asked for a likely timetable of this deterioration and was given the answer that nobody could tell for certain, but that I should expect them to fail completely in anything from six months to ten years. I was made to understand very clearly that when my kidneys finally failed, I should have to seek immediate medical assistance in the form of an artificial kidney or else surgery and a transplant.

As you might imagine, this news left me feeling despondent but there was nothing that could be done about it. After a day or two I decided to put it entirely out of mind and to live a full and normal life as long as possible. In order to do this I needed to keep my condition secret; I certainly did not want to feel that other people saw me as an invalid. I rejected the suggestion that I should have regular medical checks to keep me informed about the rate of deterioration and I also decided against any special diet or other restriction on my normal way of life. I am glad that I did this because none of what follows would have been possible if I had not and I would surely have become very poor company, as most people are when they feel sorry for themselves.

When I joined the National Museum, there was no obvious way for me to get involved with the existing administration and I therefore set about making my presence known in other ways. I engaged two men to scrub clean the exterior walls of the Museum which had accumulated considerable quantities of red dust over the past twenty years. Their wages had to be paid from the Museum Associates' coffers because the Director made clear to me that I had absolutely no authority to spend any of the money at his disposal! I recruited a gardener to help me completely re-landscape the ground in front of the Museum and I constructed some rather large bumps on the road that runs past the Museum in order to slow traffic and so help to minimize dust. The first casualty on my speed bumps was the Director, whose small car became stranded across one! My bumps were too high so we had to remake them. I was not at all popular and had to agree to keep in closer contact with the Director.

Within a month of joining the Museum staff, I had established an excellent relationship with key officials in the Government and I was able to begin formulating some new plans. In 1968, the National Museum of Kenya was nothing more than a Natural History Museum in Nairobi; the Centre of Prehistory and Palaeontology was autonomous. The various protected archaeological sites and monuments were under the administration of Kenya's National Parks. The budget for the Museum was a mere £23,000, of which Government was contributing only £17,000. I was determined to develop the Museum and to make it play a useful and effective role as a national institution.

The Government asked me to draw up a five-year plan, the first such, for the Museum. Time was very short and I had to work so fast in order to meet the deadline that I decided not to consult the Board of Trustees, but rather to complete the document on my own and submit it first to the Government. The Board continued to discuss my five-year plan for some time, not realizing the Government had already approved it. I fear I never told the Board that they were engaged in a somewhat pointless, academic exercise. I had little regard for my employers and their ultra conservative and cautious ways.

The five-year plan was to cover the period from July 1969 through to July 1974, and it gave me the chance to set out some challenging objectives against which I could test myself. In order to achieve the goals I had established, I simply had to have a sympathetic Board and I therefore asked the Minister to appoint a number of new, Kenyan, Trustees who would by their presence give Kenyans the majority vote on what had previously been a Board dominated by, generally elderly, British gentlemen.

My intention was to concentrate in the first instance on developing the Museum's educational role within Kenya. I also wanted to bring all archaeological and palaeontological matters under the administration of the National Museum. I was sure that a larger and more complex organization would have a better chance of winning Government support and I saw no reason why Kenyan taxpayers should not be directly supporting research into the country's prehistory. I personally found it unacceptable that my father was raising all the finance for his Centre for Prehistory from overseas sources. I soon discovered that this was father's way of ensuring that he and he alone had control of the Centre. Naturally enough, my efforts to bring his Centre under the Museum budget soon proved to be a serious point of contention between us, and once again I found myself at odds with my father on a matter of principle.

It was only about three months after I began as the Museum's Administrative Director that I persuaded the National Parks to transfer to the Museum the responsibility and funds for the maintenance and operation of the several archaeological sites and monuments. The effect of this transfer was the immediate doubling of my staff and a significant increase in the funds that I controlled. That it took effect so rapidly and smoothly was due to a large extent to the sympathy of the new Kenyan Director of National Parks, Perez Olindo, who had become a good friend and supporter.

The sites transferred were Fort Jesus, a 15th-century Portuguese fortress in Mombasa, and an area known as Gedi, where the ruins of an 18th-century town cover some thirty-five acres of land near the Kenya coast. The Government also agreed that the administration of all protected prehistoric sites and monuments in the country should become the responsibility of the National Museum and this of course provided excellent ground for future expansion. The Government officials concerned were very trusting, and looking back on it I am astonished that such a gamble was taken. It is a feature of young developing countries that nationals are often pushed into positions of responsibility long before they are really qualified and experienced. I was personally very happy to have proof of being treated as a Kenyan and I felt thoroughly patriotic and not in the least worried about the great task entrusted to me.

It was not long after I joined the Museum that one of my staff, who was in charge of school programmes, became involved with a number of secondary schools that were anxious to have lectures and programmes on wildlife and conservation. As a result of a seminar held in 1969, it was decided that there should be a national assocation of school youth clubs. I was approached and asked if I would help in setting up a national committee to coordinate the project, and we formed a steering committee for an organization that became known as the Wildlife Clubs of Kenya Association. By late 1969 there were something like seven clubs in Kenya and the Museum was providing a basic service to schools, which received films and literature aimed at arousing young people's interest in wildlife conservation.

It soon became clear, however, that the schools' demand for information and material was going to be more than we could possibly cope with. I therefore began to consider setting up a structure independent of the Museum designed solely to help young people develop an interest in wildlife. I had met a young American woman who was in Kenya looking for something to do. She had done some part-time work for me as a secretary in my early days at the Museum and I was aware that she wanted to stay in Kenya and do something useful. I asked her whether she would like to take on the task of coordinating the affairs of the newly formed Wildlife Clubs Association. I explained that we had no money with which we could support her but that I was sure that between us we could generate funds. This young woman, Sandy Price, accepted the challenge and before the end of 1969 she had formally accepted the post of national organizer for the Association.

I was elected Chairman of the first council of the Association and have remained so ever since. During the past decade the Wildlife Clubs movement has grown from the original few clubs to a present enrolment of over 800 clubs in Kenyan schools with an individual membership that probably now amounts to more than 50,000 boys and girls. It must be one of the most successful youth movements ever developed in the Third World and I am particularly pleased to have been associated with this wonderful organization ever since its inception.

During my trip to America early in 1969 I travelled to Washington to report to the National Geographic Society on the success of my first season at Lake Turkana. It was on this visit that I gave my first major public lecture, speaking to an audience of about 3,000 people about Lake Turkana and our discoveries in 1968. It was an exhilarating experience and since then I have always enjoyed lecturing. There is something very exciting about being on a stage and holding the full attention of a large audience. My lecture was supported by Bob Campbell's film of the 1968 expedition and I was introduced to the audience by the Chairman of the Research Committee of the National Geographic Society, Dr Leonard Carmichael.

I also used this visit to the States to request further support for my expedition and to make personal contact with a number of scientific colleagues whom I wanted to involve in future projects. It was very important that I should develop a team which had the necessary scientific credibility and I especially wanted to find people of my own age group to help me. I was concerned that every aspect of the scientific work should be undertaken by competent people. In particular, I was anxious to ensure that the fossils being collected should be placed in a well-documented geological context – there is no value in a fossil out of context.

In 1969 there were very few geologists available who were prepared to join my project on the eastern shore of Lake Turkana. To be fair, there were very few people who had experience in the sort of work that I needed done and it was not simply a matter of recruiting just any geologist. Most geologists preferred hard rocks and would avoid the detailed study of sediments that was necessary for this kind of work. Fortunately, while in America I made the acquaintance of a young American geologist who had had previous experience working in fossiliferous areas in Kenya's Rift Valley. Her name was Kay Behrensmeyer and we have worked together almost every year since then. Kay's previous experience had been at Lothagam on the western shore of Lake Turkana where she had worked with a project from Harvard's Museum of Comparative Zoology. A number of important fossils had been found there in the mid-sixties and Kay had been involved in a detailed geological study of the site. She had demonstrated her exceptional ability as well as her willingness to work in the difficult conditions of northern Kenya. Fortunately for me, Kay was not yet established and was therefore not unduly concerned at the prospect of joining forces in a new, untested project on the east side of the lake.

Altogether, my trip to the States was most successful and nobody raised any questions about my lack of paper qualifications. I was able to raise adequate funds for a second field season at the lake and I had sufficient money to be able to pay several air fares for scientists whom I wanted to involve. For the first time I was really enjoying the 'old bones' and I seemed to be establishing myself as a scientist in my own right. I was no longer just the son of the famous Louis and Mary Leakey. I returned to Kenya after a fortnight and set about planning my second expedition to Lake Turkana with great enthusiasm.

Shortly after my return from America, Margaret gave birth to a baby daughter whom we called Anna. Perhaps as a result of my excessive preoccupation with my own affairs our relationship had continued to deteriorate and later that year we separated. I did not want Anna to have the ghastly experience of living with fighting parents and I felt that the sooner the divorce went through the better. Naturally, I have tried to understand what went wrong, but the less said now, the better.

CHAPTER NINE

CAMELS AT KOOBI FORA

THE SECOND EXPEDITION to Lake Turkana was scheduled to begin in early June 1969. I had persuaded the Museum Trustees that I should be given the time to lead this project as part of my official duties. I argued that the fossils were to form a part of the collection of the National Museum and the scientific credit from the success of the project would also add to the organization's prestige. When I started work at the Museum, my opponents' major criticism was that I had no scientific experience of any kind and so I felt that a successful palaeontological mission under my direction could only help my chances of eventually winning the confidence of my colleagues.

It was as we were preparing for our second expedition that I made a major error that could have lost me vital financial support. I had been approached late in 1968 by *Life* magazine who wanted to publish a story about me. It must be understood that it was then a competitor of the *National Geographic Magazine* and, because the National Geographic Society was the principal sponsor of my project, I had to keep the story of the Turkana expedition off limits to *Life*. After lengthy correspondence explaining this to the people there, they agreed that things other than the Turkana story would form the basis of their article but at the last moment the writer and photographer were delayed and cabled to say that they could only join me late in May. I was furious because I could not delay my departure for the lake, but I was also very eager to obtain the publicity; more strokes for my ego! As might be expected, my contacts from the magazine proposed that they should accompany me to Turkana interviewing me *en route*. I was very naive and agreed on the condition that they promised to avoid writing anything that could clash with the National Geographic Society's plans for a story on my work. As it turned out, *Life* did publish a long well-illustrated article and it included a great deal about my expedition to Lake Turkana. I learned to my cost that even pleasant people have a job to do and that they ultimately place their loyalties with the source of their livelihood. Fortunately my supporters at the National Geographic Society were very tolerant and forgave me the unintentional breach in faith. Had they withdrawn their support as a result of my broken undertaking, it would have been entirely reasonable.

ABOVE The fossil search team led by Kamoya in 1975. Left to right: Tim White, John Harris, Bernard Ngeneo, Peter Njobe, James Kimani, Musau Mbitti, Harrison Mutua, Kamoya Kimeu, Wambua Mangoa and Mavudu Maluita

LEFT A portrait, photographed at Lake Turkana; I am holding an australopithecine skull

LEFT At ease with Kamoya by the lake shore

BELOW With a team of visiting Chinese scholars at Koobi Fora in 1974. Its leader, Professor Wu Ju Kang, is facing me with hands behind his back

My principal objectives for the 1969 expedition were to make a real effort to locate stone implements and archaeological sites, as well as the exposures I had found on my original helicopter trip. I did not just want to look for bones, for they do not necessarily provide any evidence of behavior. It was important, therefore, to widen the scope of our investigations so as to attempt to learn more about the life of our ancestors. The understanding of the environment in which they lived was basic and if we could find ancient living sites with tools and food debris preserved, we hoped to be better able to interpret early behaviour. I was quite certain that the area we were to work in had tremendous potential to provide the answers to the kind of questions we were asking.

The 1969 expedition consisted of fewer people than I had had with me the year before; I planned to lead a small party on our prospecting trips, while Paul Abell and some Kenyan helpers were to begin the collection of certain fossils that we had found the year before and Kay Behrensmeyer was involved with the geological studies of the fossil sites. Margaret remained in Nairobi because she felt that at such an early age our daughter should not be taken on what would be a tough expedition. Bob Campbell was busy elsewhere so we took one of the National Geographic Society's photographers instead. The other new person on the expedition was Meave Epps, a young lady whom I had hopes of encouraging to become a palaeontologist.

I was especially anxious to centralize operations in the second season and so I made plans to move our base camp northwards from Allia Bay to a spot some ten miles up the shore known as Koobi Fora. This is a dramatic sandy peninsula that juts out for nearly a mile. It first attracted my attention in 1968 when Kamoya and I were walking back to the boat after a particularly long day in the field. We had considerably lengthened our return journey by unknowingly walking all the way to the end of the spit before realizing what it was! In addition to the convenience of the superb sandy beach, I realized that a camp on the peninsula would have water on three sides, thus ensuring a constant cool breeze from the winds blowing off the water. I also felt that it would be much easier to defend a camp on the peninsula against bandits.

Koobi Fora is probably the finest and most spectacular place on the entire shoreline of Lake Turkana and I am always glad that I made the decision to camp there in 1969. To get there I had to forge a track for the vehicles but by good fortune we found a way that proved relatively easy and it still serves today as the main route into Koobi Fora. It took only a few days to establish our camp, which consisted of a few tents and a single grass hut that was used as a laboratory. We were also able to mark out an airstrip on the mudflats of a small dried-up lake some six miles from our camp site. This was useful because I had to make several visits to Nairobi.

I again had guards with the expedition, but on this occasion only two men instead of the previous thirty-eight. I was sure that we were unlikely to be attacked by the bandits whom, I guessed, were more interested in cattle than a group of rather strange scientists. During the previous expedition we had not seen a soul and I felt that the possible risk was more than offset by the tremedous advantage in having a smaller party. The one major drawback to Koobi Fora was that drinking water had to be fetched from Nderati or some other water hole usually more than fifteen miles distant over rough tracks, so the fewer of us there were the better! I became so used to the need to conserve water at Koobi Fora that I still worry every time I see water being wasted; even in places where there is no shortage.

One of my first objectives was to get Kay started by showing her the area that we had covered in the previous season. She had two basic tasks: first, to draw up a geological framework which would help us to place the fossils in their correct context and establish their relationships to one another, and second to locate horizons of volcanic material that could be used for dating. Kay had not had the advantages of a previous season in the area therefore she had to cover a lot of ground quickly. The dating of the site was critical: we did have a general idea of the age of the fossils from comparisons with Olduvai and the Omo, but it is always preferable to have precise dates. To achieve this Kay needed to find outcrops of volcanic ash suitable for dating, interbedded with the layers of silt and sand containing the fossils.

These volcanic ashes, or tuffs as they are known, often contain lumps of pumice full of minerals which were crystallized at the time they were spewed out of the volcano. It is the crystals rich in potassium that are used in the method known as potassium-argon dating. I had arranged for Dr John Miller, a British geophysicist at Cambridge, to receive our samples as soon as we could get them to England. Miller had agreed with his colleague Frank Fitch to run our dates, and our understanding was that this would be a straightforward commercial undertaking for which we would pay them. At that time they were operating a dating consultancy known as F.M. Consultants Ltd and I was to be, at least in the initial stages, simply a customer. Of course, this was not ideal and I would have preferred a relationship based solely upon scientific interest, but at that time I was pleased and quite satisfied to know that Miller, then considered by many to be Britain's leading expert in dating, was working on my project. Later on, as I shall describe, the dating programme at Lake Turkana became extremely controversial and my relationship with F.M. Consultants changed.

As it happened, our preoccupation with dating material led to an unexpected find when Kay was out collecting pumice samples. At one particular locality Kay noticed that there were a number of lava flakes lying on the surface of the eroding tuff where she was working. It was during the first few weeks of the expedition and I had actually returned to Nairobi briefly because I was needed for a Museum meeting. Kay recognized that her discovery had to be checked as she realized that the very simple flakes could be the first record of early stone tools in the vast area we were searching. She returned to Koobi Fora and decided to do nothing further until I returned to camp.

When I arrived back at the lake late in the evening accompanied by Meave, I was tired from the long twenty-five-hour drive which we had completed without a break. All the same, Kay's news was very exciting and we agreed to go to the site first light the next morning. I confess to having been somewhat sceptical and probably very jealous: how could a geologist recently introduced to the area find something that I had looked for unsuccessfully during a period of about four months! I consoled myself with the uncharitable thought that Kay was almost certainly mistaken; she had probably found recent stone artefacts of which there were many in the area and in which we had no interest.

The next morning we set off early to make the two-hour drive to a point from which we could walk to the site. As soon as we arrived, there was no doubt in my mind, Kay had indeed found early stone tools, and what was especially exciting was the fact that they were apparently eroding from a tuff which could be dated. I was tremendously pleased, and while standing at Kay's site I suddenly realized that in the previous year's exploration I had come across an almost identical outcrop of tuff. My subconscious was telling me that I had seen the same thing almost a year before. Could tools be there too that I had simply missed? It is quite easy not to see things that are later obvious and I am convinced that I search with a predetermined image in my subconscious. If I am searching for fossil bones it is unlikely that I will find stone tools and vice versa. Anyway, I decided to allow my subconscious to lead me back and I excused myself from the party, saying that I wanted to have a look around on my own.

After about half an hour and a mile of walking through the badlands I came to exactly the place I had remembered and there to my delight were stone flakes too, just like those at Kay's site! I was ecstatic and rushed back with my news. It was necessary to excavate one of the sites to prove that the stones were really coming out of the tuff and were contemporary with it. We agreed to dig a trench at Kay's site. After we had completed a small excavation over several days we found more flakes *in situ* and thus proved that they were contemporary with the ash and therefore very old. We chose not to collect anything at the second site which I had found because I wanted to leave the surface evidence intact for whoever was going to work the site in future.

This discovery of ancient tools provided an added urgency to the dating of the tuff. The tools were from a level that had yielded various fossil animals and these had implied a probable age at least equivalent to the oldest horizons at Olduvai, the base of which was 1.9 million years old. At that time we were all hoping that the dating would establish Kay's find as the oldest evidence for technology from anywhere in the world. Within a matter of weeks we received a response from Fitch and Miller. Provisional interpretation on the samples we had sent gave a date of 2.4 million years, but they wanted further samples of the pumice to confirm the age. As you might imagine, we were beside ourselves with joy; this date suggested that Kay's tuff and its associated artefacts were almost half a million years older than anything that had been found before.

In a relatively short space of time we were referring to this geological horizon as the Kay Behrensmeyer Site tuff, hence the now well-known term, the KBS tuff, which subsequently became the focus of a raging debate. In 1969 we were ourselves anxious to have confirmation of the provisional date and so Kay collected a number of additional samples of pumice which were sent to England for analysis by Fitch and Miller. The date was not only supported by these subsequent samples but a refinement in the procedure suggested that a date of 2.6 million years was more accurate.

As it happens the date was wrong and the KBS was to become the central issue in a saga that lasted well over a decade. Eventually, its age was recalculated to 1.89 million years and in a later chapter I shall describe some of the battles which took place. In 1969, however, we had no reason to doubt Fitch and Miller who had dated many East African sites and who enjoyed an excellent reputation in their field.

With the discovery of the tools I was especially eager to find the fossil remains of the hominid who might have made them. This was a crucial matter. For a long time it seems to have been generally agreed that the first making of stone tools is a sign that our ancestors had finally emerged as humans from their animal background. What we really wanted to know, however, was which hominid form had crossed this threshold? Was it an australopithecine, *Homo habilis* or some new species of *Homo*? Who were the first tool-making humans? In order to answer these and similar questions I needed to explore the area thoroughly, and for this purpose I decided to use some camels.

I made the necessary arrangements and acquired several of these beasts which, I reasoned, would enable us to get to places that would have been difficult to reach by Land-Rover and at the same time they could carry enough food and water for a week or more. It was a romantic notion and I have to admit that I rather fancied my image as a camel-riding explorer of the great African desert. Our camels included four retired police-riding camels, which we borrowed, and six other animals which we rented from a small village called Maikona, about 150 miles away from the lake. It was extremely difficult to buy camels so I arranged the rental at a daily rate which today would be equivalent to $6.00 for the lot. Camels take a great deal of understanding as well as encouragement and because I had absolutely no idea of where to begin, we also hired two camel attendants. On their way to the rendezvous we had arranged near Allia Bay, the camel party was ambushed by bandits at a water hole and had to take cover. Shots were exchanged for several hours but fortunately none of our people was injured although one of the bandit attackers was killed. A pride of lions also attempted to eat our camels but despite everything all the animals escaped and arrived safely after six days of travel. They were established within a small thorn-fenced area near the camp which served as their 'stable' at night. As it happened, our fence was quite inadequate and whenever lions came near we had to go out and protect the camels.

The easiest part of camel riding is thinking about it before hand. The first thing we had to do was to learn how to ride a camel. My team was to consist of four riders: Kamoya, Peter Nzube, Meave and myself. Three camel attendants were to lead the six pack animals with our water and provisions. None of us had ridden a camel before and Kamoya and Nzube had never ridden anything without wheels! Camels are taller than horses and their humps make it impossible to use anything except a specially designed camel saddle. There is no bit for the bridle, but instead the reins are attached to a halter and can be used to pull the camel's head one way or another. Fortunately a camel has a very long weak neck and it is reasonably easy to pull the head in the required direction. Stopping and starting as well as changes in speed are best initiated with spoken commands. To begin with we had problems because nobody had told us this and, anyway, we had no idea what sounds were likely to be understood by the beasts.

The first hurdle was to saddle the camels, and it took us a while to discover how to do this. The obvious difficulty was that a large riding camel is much too high for a person to stand alongside with a saddle. We tried all sorts of devices, all of which ended in disaster. The wretched animals were becoming quite distressed, as were we, when one of the camel herders came and said something to one of the animals and to our surprise it immediately crouched down. After this it was easy and we learned that one has to ask a camel to sit down to saddle it and generally it will. How different from a horse! Camels are noisy and they usually keep up a deafening bellow while being loaded, swinging their heads about, spraying a nasty smelling green saliva and bits of partially digested food all about them.

Having finally saddled our animals, we were naturally eager to mount, and, while the animals were down, it looked as though it would be easy. It is indeed easy to get on, but a camel has an unfortunate way of getting up: it raises its rear end first, following a few moments later with the front. The effect of this was that we were immediately pitched off! There were hilarious moments, but after an hour or so, the four of us had found a system that seemed to work.

Although riding a camel through thorn bush is probably one of the least satisfactory ways of getting about, I have some extremely happy memories of our trips. As a general rule we took an absolute minimum of equipment but as much water and provisions as the animals could carry, a weight of about 200 lb. The loads were split into two and attached either side of the hump. Our animals seemed to be unhappy all the time, but they became much more bad tempered the longer they were without water. We tried to let them drink at least once a week, although I was told that they could last for two weeks without any serious problems. If one can accept the discomfort and natural idiosyncrasies of the camels, they provide a marvellous way to travel through wild unspoiled country. Progress is always slow and one has plenty of time to see the countryside.

Our normal routine with camels was to begin the day just as the dawn began to lighten the eastern horizon. One person made a fire while others busied themselves with preparing loads. Tea was drunk sitting round the fire in the half-light and well before the sun rose all hands were busy with the loading. Generally we worked in pairs, each of us having responsibility for certain loads and particular animals. The first step was to grab the protesting camel's upper lip so that the second person could fix a rope to the lower jaw. While this sounds easy it really takes a lot of courage because these animals have sharp teeth and will readily bite to prevent being bridled. Once the lower jaw is tied, however, the animals recognize your authority.

The loads for the camels were attached to a frame of four sticks, two on either side of the animal's body. These were joined below by means of two sets of ropes that passed under the chest and further back below the beast's belly, while the upper ends were firmly tied and padded with hides and sacking. The loads included a variety of boxes, cartons, pots, pans and blankets which were tied with lengths of rope. It was amusing to see just how many different items of varying shapes and sizes could be tied on by a skilled person. Once every animal was ready with all the loads secure, the command to stand was given and all the camels would lurch to their feet.

It was usually at this point that the fun began because quite often a load would come loose as the animal struggled to its feet. The noise of a box full of equipment such as metal plates, mugs and cutlery crashing to the ground invariably produced a stampede, and within a few moments there would be camels in every direction with their loads strewn all over the place. Whoever had loaded the animal that caused the stampede would always come in for a good deal of abuse, providing added incentive to tie the loads on properly in the first place. Once under way, there were often further problems with the ropes working loose. In spite of these inevitable delays we could keep up a marching pace that I estimated to be three miles an hour.

The camels had to feed for several hours a day and we used to stop travelling at about eleven, leaving at least four hours for the animals to browse before continuing our journey in the cool of the late afternoon. Because of the many lions in the area, I always halted at the end of the day with sufficient time to unload and hobble the camels before nightfall. As a rule we attempted to find a large area of clear ground in which to spend the night so that prowling animals could not get too close without being seen. The camels slept in a tight circle, each animal facing outwards. We arranged ourselves to sleep in a wider circle around the camels thus providing a human barrier to any prowling lions. This also meant that we were strategically placed to grab any camel that attempted to bolt. Even though hobbled, camels can show a surprising turn of speed when frightened.

Generally we ate very little or nothing during the day; we drank tea at dawn and sometimes at noon we would cook some potatoes and dried meat. Our main preoccupation was thirst; we drank on average five litres per person during the course of the day. I would usually prepare the evening meal for the group: a meat and vegetable stew or curry with rice. At the end of a long day everybody had quite an appetite and a large meal was welcome. After dinner, we drank coffee while sitting round the dying fire and this was the time to talk over the day's events as well as the world's problems. I find it difficult to describe the feelings of content that were generated by these evenings under the brilliant African starlit skies, listening to the rustle of the wind and the distant calls of jackals, hyenas and sometimes lions. To be sleeping on the warm ground without any shelter is perhaps as close as one can get to the conditions under which the people of our prehistoric past must have lived.

One of my main worries at night was that we would be bitten by snakes or stung by scorpions but although we saw plenty nobody got a bite and nobody stayed awake worrying. During the night the camels were often very restless; I would awaken to loud snorting, which was their alarm call, and on checking, found that all the animals would be staring in a particular direction. This usually meant that a lion was nearby and we would stoke up the fire, keeping watch until the animals relaxed again. Had the lions attacked, we were powerless to defend the camels but fortunately this never actually happened. There were also many occasions when the camels would be startled by something and we would wake to see them hopping away into the night on their three free legs.

Our first camel trip lasted about a week and we soon discovered just how difficult it was to operate with them. As with horses, the most difficult position is in front because the others will tend to follow, and it was a particular challenge to convey to my mount the fact that I was the leader. The physical effort required to induce a camel to go where you want it has to be experienced to be appreciated. In addition, the extraordinary swaying movements of a camel resulted in many strained muscles and by the end of each day we were absolute wrecks. Quite apart from these obvious problems, the camels that we were riding would take every opportunity to brush us off their backs by walking under low hanging thorn bushes. Sometimes they would unilaterally choose to stop and rest and no amount of effort would get them going again until they were ready. All in all it was most unsatisfactory and the end of the day was looked forward to with eager anticipation. It was not at all romantic, and looking back I am quite unable to understand why we made several trips on them. We eventually decided to use the animals only for baggage, while we walked in preference to the torture of riding. In fact, a walking journey with camels can be a great deal of fun.

Before learning this lesson we embarked on our second camel journey and it was on this trip that Meave and I made our first discovery. We had travelled north from Koobi Fora about fifteen miles and were heading towards an area of sediments near the Kenya-Ethiopia border. On the third day out, the now familiar backache and other discomforts from riding the camels had returned and I was looking for every opportunity to rest the convoy. Quite early in the afternoon, several hours before our normal stopping time, I saw some small exposures of sediment which offered a chance for us to explore on foot. I called a halt and we turned the camels over to the attendants, promising that we would be back before dusk to help settle the animals for the night.

As it happened there were quite a lot of fossils in the area and so we decided to spend at least another morning searching before moving on. The next day, 27 July 1969, the four of us, who were prospecting in pairs – Kamoya and Nzube working in one direction while Meave and I planned to search in another – agreed to rendezvous at one o'clock at camp. It was a very hot day and by about 10.30 a.m. I had had quite enough. We had seen less and less evidence to suggest that the area warranted extensive searching and I was quietly convincing myself that I could justify an early return to the camels where there would be a chance to have a cool drink and rest in the welcome shade. I did not want to set a bad example and I certainly did not wish Meave to realize that I was getting soft and so I pretended to be feeling unwell. I opted for a stomach pain and backache and Meave quickly proposed that we should return to the camp. The decision to quit was hers! I was delighted and we set off to retrace our footsteps, eager to reach the shade and a drink as soon as possible.

I took the lead choosing the quickest route. I decided to walk down a small dry stream bed. The speed with which I headed back should have given away my duplicity but before long our progress was unexpectedly halted. I was leading, several paces ahead of Meave, when to my astonishment I saw an *Australopithecus* face looking at me from a distance of about fifteen feet. I instantly recognized what turned out to be a complete fossil skull lying on the sand.

You can imagine our feelings. The skull was readily recognizable because it was remarkably similar to 'Zinj', the *Australopithecus boisei* skull found by my mother at Olduvai in July 1959, almost ten years to the day before our discovery. At that time, this had been the only skull of this species found in East Africa, and as her find had been fragmentary it was several weeks before enough reconstruction had been completed to see the 'face' of the skull. Yet here were we, staring at a complete skull in the bed of a sand river, and for a moment or two I was sure that it was all an illusion brought on by the sun. We crept forward, somehow afraid that the fossil would vanish before our eyes! It did no such thing and in a few moments we were peering at it from just a few inches, marvelling at its completeness and the incredible fact that *we* had found it. It is the only time that a discovery has left me truly breathless.

Meave and I sat under a small thorn bush trying to keep cool while keeping the skull under constant observation. It was such a dramatic find that I wanted Kamoya and Nzube to share in our excitement and to see it as we had, lying in the bed of the stream. Our rendezvous was about a mile away and at least an hour off. Could we leave the skull for an hour unattended or should one of us remain on guard? I was reluctant to leave Meave alone and she was reluctant to return to the camels alone. Consequently we agreed to set off together, but we built three enormous stone cairns on high points around the site, making sure that we would not fail to find the specimen when we returned.

Kamoya and Nzube arrived back at the camp at about 12.30, having had an uneventful morning. It was incredibly hot and still. Normally, this time of day is best spent in the shade, but on that occasion, as soon as my friends had had a drink, we all set out again almost running. I was greatly relieved and slightly surprised to find the skull again and I took tremendous pleasure in pointing it out from a distance. Both Kamoya and Nzube were so elated that they lifted me shoulder high and did a sort of dance of joy for a few moments. None of us noticed the heat and we spent a lot of time just sitting near the skull looking at it and chattering excitedly about our good fortune. At last some sense returned and we agreed to go back to the camels and our camp for a rest. We collected the skull in the late afternoon when it was cooler.

It was an easy specimen to collect because there were no broken fragments. The skull had washed out from the sediment as the stream cut back the bank during the previous rains. The natural impression of the fossil was plainly visible in the bank. It is certain that it would have been washed away or badly broken in the next rains. The only parts that were missing were the teeth and I believe that these had been broken off and lost before the bone became fossilized. Usually skulls are broken up before fossilization and it is extremely uncommon to recover such a complete specimen.

Back at our camp that evening, we had a discussion about our next move and it was agreed that we should return at once to Koobi Fora. I was anxious to get the skull to safety and it seemed unwise to continue northwards on our camel trip with such a precious cargo. I was also eager to contact Kay so that she could study the site and determine the skull's position within the overall geological section. And, of course, we all wanted to share our news with the others at Koobi Fora. We set off next morning at dawn hoping to make it back to camp in two days. The skull was carefully wrapped in sheepskin and placed in a small box which was roped firmly on to my saddle.

On our journey from the north with the skull we went at a good pace and arrived at Koobi Fora late on the second day. I was extremely pleased to find that my mother was in camp, having flown in on her way back from a visit to the Omo with Clark Howell's expedition. She had been working at Olduvai so I had seen little of her for several months. We had sent word to my mother following Kay's discovery of the tools and she had taken the opportunity to see the new site for herself. We had known of her plans but as we expected to be away on the camels, we had left the arrangements for her visit to Kay and the others. Because we were not expected our arrival caused considerable consternation. The sight of a group of people bearing down on the camp at a fast pace was very alarming, until we were recognized!

It took very little time to unload the animals and we were soon ready to show our find. It is difficult to describe the surprise and delight of everyone when I carefully lifted the complete skull from its box and placed it gently in my mother's hands. It was the sort of moment in one's life that can never be forgotten. I had as much pleasure from my mother's reaction as I had from finding the skull itself! We sat around the table discussing it for a long time. Naturally my mother was particularly anxious to compare it with her own find of ten years before but, unfortunately, I did not have any casts of '*Zinjanthropus*' or other fossil hominids at Koobi Fora. Because I am superstitious I felt sure that casts would have brought bad luck to our search efforts. It is only recently that I have relaxed this rule and now we do have casts at Koobi Fora.

My mother left for Nairobi the next day and we prepared to return again to the new skull site for follow-up work. I decided that we should go back with the camels so that afterwards we could continue exploring. We also arranged for a vehicle to meet us at the site so that extra people and supplies could be brought in. Several days later we all met at the appointed rendezvous and a small camp was established. I wanted to sieve the dry river bed around the place where the skull had been found just in case the missing teeth were in the sand. As it turned out, the teeth were never recovered and our sieving operation produced absolutely nothing.

While camped at this site, I decided that we should spend a little more time on searching for other fossils. On the second morning, Meave and I were working quietly on the sieving when we heard Kamoya, Nzube, and another colleague, Mwongella, returning to us talking excitedly. It was several hours before lunch and their unscheduled return could only mean a discovery. Mwongella had found another human-like skull about a mile away. It was hard to believe their news but we rushed off to check. It was, as they said, a hominid skull but this time less complete, constituting only the back and base of the cranium; the eyebrows, face and upper jaw had broken off and, as the breaks were not fresh, it is probable that they must have been lost some time before the find was made. It was quite extraordinary to have such luck and I realized then that a careful search of the whole area would almost certainly yield many more specimens. So began my planning for a really long-term project. I resolved that I would make a major commitment to organize a large interdisciplinary effort in which a variety of scientists would work over a long period of time to study in full this rich fossil site.

After about a week we had completed the sieving at the first skull site and I had collected all the obvious surface fragments of the second skull. Kay had completed her preliminary study of the geology and so we went on with our exploration of the area to the north. We went up to the Ethiopian border and then marched westwards along the Kenyan side of it to the lake shore. The principal conclusion I reached after this trip was that on my reconnaissance flight two years before my helicopter had never got into Kenya. The stone tools and fossils that I had originally seen were in Ethiopian territory, and so out of reach of our expedition! To this day, these sites across the border and south of the Omo have not been studied and will, in my opinion, offer some future archaeologist considerable joy.

As our expedition progressed it became very clear that in future we would be spending a great deal of time along the eastern shore of the lake and I turned my thoughts to the problem of accommodation. The tents that we had were suffering very badly from the incessant wind and they were not particularly comfortable for working in. I decided that the following year I would build a permanent camp of grass huts of various sizes. I also had the problem of what to do with the rapidly growing quantity of field equipment that was simply too bulky to move regularly between Nairobi and the lake. The solution seemed to be to construct stores at Koobi Fora where things could be left between expeditions.

Every day, during the time that we were at Koobi Fora in 1969, when we returned from the fossil exposures Meave and I used to collect half a dozen or so large flat stones. These are abundant in the area and often contain fossilized snails or animal bones. Each evening we took the stones to a spot on the Koobi Fora spit which I had selected to build our permanent camp. Gradually we laid out a stone floor that measured 50 × 15 feet. Before the end of the expedition I arranged to buy and transport two prefabricated metal huts, known as Uniports, which we erected on this floor for the stores. When it was time for the expedition to go back to Nairobi, we left a lot of the non-perishable material in one of these huts. After all, we had seen no sign of any people in the area for two successive three-month seasons.

I also decided that in future it would be more economical to bring heavy supplies such as fuel and bulk foods from Nairobi via the west coast where the access road was so much better. For this I needed a large boat which could be used as a ferry between the small fishing village situated at Kalakol (then known as Ferguson's Gulf) and Koobi Fora, a distance of about thirty-five miles. We already had a boat but this was rather too small so I bought a new and larger boat. By October this new motor launch was ready for use and Meave and I decided to go to Koobi Fora via Kalakol for a long weekend to see how our stores were. This was the first of several such visits which we made despite the tremendous distances involved and the bad roads which made travel slow.

We drove to Kalakol overnight and set off for Koobi Fora the following noon in our new boat. We arrived at the camp and found everything was just as we had left it, secure in the locked store; there was no sign that anyone had been anywhere near it. We had the place entirely to ourselves, not another human for miles and miles, or so we thought. On the following morning, we decided to go for a long walk along the beach and on our return I suddenly realized that the footprints on the sand could not be ours because they were leading towards, rather than away from camp. Some people had come along the beach while we were away and were probably in camp helping themselves from our unlocked building. I was a little alarmed because we were quite unarmed and all the people who came through this area of Kenya traditionally carry rifles. If they had any bad intentions we were certainly in no position to argue. We arrived back to find nothing harmed and the tracks showed that a group of about five men had come up the beach, passed through the camp and then gone on south. Was it a ruse? Were they waiting in hiding to see if we were armed? Needless to say, the incident quite ruined the weekend and we decided to set off back across the lake to Kalakol after relocking the store.

Naturally this first encounter with unseen local people gave me cause for concern. I had thought that the place was completely uninhabited and that our stores would therefore be quite safe. Now I was not so sure. We returned a few weeks later to check once more. Meave and I drove to Kalakol on a Friday night and upon reaching the camp the following afternoon, my worst fears were confirmed. The metal door of the hut had been forced open and the contents of several tin boxes were strewn about all over the floor. Several blankets had been stolen but there were no serious losses. That weekend we saw nobody but we spent a lot of time burying the metal boxes containing the crockery, blankets and various utensils in the sand outside the huts. I believed that our visitors would not really disturb the large canvas tents, the beds and other heavy gear but would be more likely to make off with blankets, cups, bowls and other small but useful articles. When we had finished, everything was well hidden. This time I left the hut unlocked so that any visitor could see without difficulty that there was nothing to take and there would be no need to smash the door or windows.

We returned at Christmas to find that our strategy had worked well and I was confident that everything would be fine until the start of the next season in June 1970. How wrong I was; when Meave and I returned to Kenya in February after a visit to the U.S.A., it was to find a message from the local National Parks Warden. He had called in at Koobi Fora and had found everything vandalized. We went there as soon as possible and it was as bad as we had been told, everything that could be broken, cut, torn or unwound had been – I have never seen such a complete scene of destruction. What seemed to have happened was that a large party of men had spent a full day in the shade of our huts busily examining every artefact of Western technology. Films had been unwound right down to the spool, tape measures pulled out of their containers to the last inch, buckles ripped off, ground sheets torn from the tents and removed, covers ripped from mattresses and the foam fillings torn. Ropes had been removed from the tents and even the bed springs were gone and were later seen worn as earrings or necklaces on the local people! Pots, pans, buckets, jugs and tins of food were all missing. The only remaining food was an evil-smelling sardine tin which had been badly opened. There was lavatory paper everywhere and all the items that remained were jumbled together in a huge pile in the middle of the floor. Pots of glue, ink, varnish and paint had been emptied onto this pile making it even more difficult to sort out. It took us several hours with frequent bathes in the lake to cool off before we managed to restore some sort of order.

Later on we learned that it was a well-armed party from the Ethiopian side of the border which had gone over our stores, and on reflection I can hardly blame them. It was we who were naive in leaving such tempting articles unattended in an unlocked store. In any event there was nothing to do but re-equip for the 1970 season. Since that incident Koobi Fora has never been left unattended.

CHAPTER TEN

MAKING CHANGES

IT WAS IN THE PERIOD between the 1969 and 1970 Koobi Fora seasons that the Director of the Museum departed. As no senior scientist was appointed to replace him, I quietly assumed complete control under the supervision of the Museum's Board of Trustees. This was what I had really wanted: the chance to develop Kenya's National Museum according to my own views. The challenges were clearly going to be easier to deal with as a result of my increasingly independent identity as leader of the Koobi Fora project which had already focused considerable international scientific attention on Kenya and on me as an individual.

One of the very first things that I had to contend with was the problem of the autonomy of father's Centre for Prehistory and Palaeontology. The fossils and stone tools from Koobi Fora were to be studied within the Centre, where I had no authority, the staff there being directly responsible to my father. I set about the task of persuading father and others that this arrangement should be changed. I felt that there should be just one administration and that it should come under local control and be financed by the Kenya government. By early 1970 my father was spending a great deal of time overseas seeking funds for various projects and as a consequence both his scientific work and his health began to suffer. He was concerned with more than just Olduvai, where my mother continued to work, and the Centre; he had become deeply involved with helping to raise support for various primate studies in Africa and archaeological excavations beyond Africa. The field study of chimpanzees by Jane Goodall and later work on the Mountain Gorilla by Dian Fossey were both projects of great importance to him. Indeed, both projects were conceived and set up by him. In addition he was involved in the development of a small colony of captive monkeys which he saw as the nucleus of a primate research centre in Kenya. This was the centre at Tigoni where Meave had worked when she first came to Kenya. A project in Israel was also taking his time and he had taken on commitments to an expensive and controversial archaeological project in the Mojave desert of southern California of which many of his colleagues, as well as my mother, were distinctly doubtful. My father's efforts to find early man in California have been described by Sonia Cole in her book, *Leakey's Luck*. His reputation

ABOVE With Louise, Samira and Meave at Koobi Fora

LEFT Anna, my first daughter, and Louise pouring water over me. The camp at Koobi Fora forms the horizon

Meave, Louise and Samira in local Kenyan dress,
known as Uitenges, at Koobi Fora

probably suffered in certain circles as a result of his work there. In contrast was the very important project at 'Ubeidiya, which lies in the Jordan valley, between Israel and Jordan. The site is situated in a northern continuation of the Great Rift Valley that has yielded so much of value further south. The Dead Sea, some miles south of 'Ubeidiya, is a rift-valley lake very much like Natron or Turkana. The sixty or more archaeological levels of the 'Ubeidiya site contain tremendous numbers of stone tools and some animal fossils, but no useful hominid fossils have been found there to date. The site was discovered in the 1960s and my father was asked by the Israelis for his advice and later for help in raising funds. He assisted them gladly but it was another time-consuming commitment in yet another area, which detracted from his work on prehistory in East Africa.

It was against this background that I put forward an idea which would lead to the eventual transfer of financial responsibility for the Centre for Prehistory to the Museum budget. I thought that father would welcome the release from one of his major financial burdens. The idea that Kenya should support the work that he had pioneered and financed alone for so long would, I hoped, appeal to him. In fact, it didn't, because he thought that the Kenyan Government would not be able to continue to support the project once it was started. My discussions with officials in the Government indicated a willingness to take on the extra costs provided that there was a centralization of budgets under the overall control of the Board of Trustees. My ideas were also supported by the Board, especially after the appointment of a Kenyan Chairman to replace the British Chairman, who had known and worked with father for so many years. It took a number of discussions to convince everyone but by early 1970 I had included a budget provision for a part of the Centre's expenses in the overall submission to the Kenya Treasury. I had no intention of changing any aspect of the Centre's scientific policy and agreed readily that my father should continue to direct the affairs and staff as he had always done.

For reasons that I do not fully understand even now, my father began to feel and believe that I was actually working to eliminate him from any participation and involvement. I suppose he was bound to feel put out because I wanted to do things differently from the way he had always done them. He became deeply suspicious of me, and for a while every action I took was open to misinterpretation. It was thus that father and son were seen to be at odds and people began to take sides. Father had many friends abroad, particularly in the U.S.A., where each year he spent increasing amounts of time raising funds and lecturing at colleges and universities.

His friends helped in his fund raising and they must have thought that I was a young upstart with no respect for the great man. He became particularly involved with a group in California where some of his friends and supporters had established the L.S.B. Leakey Foundation. This organization relied heavily on father's charismatic personality which at that time was the main thrust in its fund-raising efforts. Because of the name and endeavours of the Foundation, father felt entitled to request that much of the Foundation support be put towards various projects in which he was interested.

I was also beginning to visit the U.S.A. each year and my efforts were directed to raising support for my own project at Koobi Fora. Increasingly I too was being invited to talk at colleges and this, together with the spectacular fossil discoveries at Turkana, resulted in a great deal of publicity for me. The United States is a big country and there was, in fact, no real competition between father and son in terms of meeting the right contacts for fund raising. Nevertheless, it soon became a point of conflict between myself and a number of father's devoted supporters in California. While my own relationship with father had become increasingly strained, this was nothing compared to the animosity that some of my father's friends developed towards me.

In Kenya everything was going well. I was beginning to understand what museum work was all about. I had high hopes that the Museum would receive greatly increased financial support from 1 July 1970 in response to the five-year plan that I had submitted. I had proposed that between 1970 and 1975 there should be an expenditure on development of more than £850,000. It was a preposterously high budget when put against budgets in previous years but I was not particularly concerned with such comparisons.

This dramatic increase in proposed Government funding was justified in my five-year plan by the explanation I gave of the role a museum should play in a developing country. I still believe that museums have the potential in young countries of shaping a sense of national identity, although sadly this is seldom realized. All too often, the countries that were former European colonies now have museums that are simply reflections of colonial times when private interest was the main focus in a museum. The attitude of governments in Third World countries so often mirrors this. A national treasury will forever be seeking ways of not spending money and it takes considerable efforts to persuade the officials within a large bureaucracy to adopt new attitudes towards non-revenue-earning projects.

I believe that a museum system can be designed to play a significant national education role. Through exhibits and special programmes it is possible to stimulate interest in science and pride in a national cultural identity. To achieve this a national museum must be extended into the rural areas where there are large numbers of people. In Kenya it was obvious that a museum in the capital could never serve more than a small fraction of the total population. The Nairobi Museum had been established to serve only the city's inhabitants and a few visitors.

My plans to extend the Museum's activities and influence beyond Nairobi received a marvellous boost from an unexpected bequest. On his death, a farmer in Kenya's western province, Colonel Hugh Stoneham, who had been operating a private museum for many years, left his estate, which included a valuable library, to the country. The executors of the bequest agreed that the Stoneham farm should be sold and the proceeds used to build a new small museum. I made arrangements for the acquisition of a suitable plot of land in the township of Kitale.

The new Kitale Museum was the first of a series of regional museums that I hoped would be built in Kenya. I felt that these museums should serve the communities of the respective regions and although they would remain constitutent parts of the National Museum, I wanted to create a strong local focus of interest. To achieve this, various leaders of the Kitale community were invited to make up an Advisory Committee which would work closely with the Museum Curator and staff. These regional museums were seen as educational tools and not research institutions; I have attempted to focus all museum research at the Nairobi headquarters. There was considerable need in 1969–70 to expand the direct educational role of the Nairobi Museum and I made a deliberate choice to spend more in this area at the expense of various scientific activities. The one exception that I made was, of course, the research in prehistory. There is no doubt that my own interests at Turkana had an influence on this and in recent years I have given absolute priority to prehistory and the need for adequate research facilities. Although much had been done over the years, the actual laboratories and storage facilities in Nairobi were totally inadequate and something better had to be done. I was also very anxious that more time and effort should be spent on the preservation of ancient monuments, the relics of ancient towns, that are scattered along Kenya's sea coast. Also, better facilities were required for the natural sciences, but I made this my second priority.

One unusual but sometimes difficult problem that as Director I had, was to establish my authority with a number of staff who had worked for the Museum longer than I had been alive! Because of my father's position at the Museum many of the staff had known me since early childhood and, inevitably, I worried a lot about this. Was I the real boss, or did people simply support me because I was my father's son? Certainly he had many very loyal supporters, especially among the Kenyans who felt that he was one of them; a fellow citizen who understood the ways of the country.

At this time in Kenya's history there was considerable resentment of colonial attitudes of superiority. An intriguing aspect of this is that this resentment was not directed at any particular racial group. However, sadly, it was mainly immigrant Kenyans who would not or could not adapt to new conditions. Racial prejudice is largely a characteristic of the non-African communities in Africa. It is astonishing to me that any immigrant groups were permitted to stay on in the post-independence period when one considers the extent to which the colonials abused and exploited the indigenous people. There are still people who live in Kenya who have absolutely no respect for the new nation nor do they identify themselves with its populace. This group has been allowed to remain because of its purported contributions to the country's economy. The expulsion of a member of this very privileged group immediately results in a very hostile Western press which otherwise tends to ignore African events.

In 1969 Margaret and I finally parted company. She stayed on in Kenya and took up a position in the Centre for Prehistory where she initiated work on a national catalogue of all the specimens from various palaeontological sites in the country. As a result of her work, I learned of the great quantity of important Kenyan specimens which were then abroad being studied by foreigners. At that time, the Government of Kenya was quite willing to grant temporary export permits for scientific material, allowing it to be studied abroad, and there was little pressure for its speedy return. I felt that there is very little value to a scientific collection that is split up and not readily accessible to scientists in one place so I began to request that Kenyan fossils be returned. To be fair, there was not enough room in the country's museum to store a large collection and I was soon under pressure to stop my campaign.

During 1970 a number of ideas began to mature in respect of the work at Lake Turkana. Our discoveries during the 1969 season had been both spectacular and important. It was clear that numerous scientists would need many years to study fully the vast site which we knew covered at least 500 square miles. By comparison, the Olduvai site was less than 35 square miles while the Plio-Pleistocene fossil locality at the Omo covered about 105 square miles. It now became obvious to me that the adventure aspect of the expedition which I so enjoyed was being replaced by the responsibility of mounting a serious and competent scientific project. This was exactly the challenge that I had hoped for when I opted out of the Omo Valley expedition because it provided me with the unique chance to do things independently. I would, I hoped, establish my own reputation in the field and, if successful, be accepted by the professionals in my own right.

I saw the need to build up a strong team of scientists who would be prepared to work with me for a number of years on the project. Unlike other sciences, prehistoric studies are still inadequately financed and I am not in a position to be able to hire scientists on a full-time basis. Instead, I look for people who are prepared to work with me at Koobi Fora for several months each year over a number of years without any reward except for the privilege of being involved. I also look for scientists who can attract financial support for their own work from the various sources that are open to European and American workers. It was in this way that the project grew into one of the largest interdisciplinary, multinational teams ever assembled to investigate a specific prehistoric site in Africa.

The problem was to decide whom I should involve. I needed people to study the various fossil groups, especially the hominids, but I also required archaeologists, geologists, palaeoecologists, geochemists and geophysicists. I myself was not qualified to take on any aspect of the research, except perhaps the description and study of any fossil monkeys that we found. After a lot of careful thought I decided that I would not only invite the participation of scientists who had been involved in research in prehistory in East Africa, but I would also include scientists with experience elsewhere. I was especially impressed at the time by Clark Howell's concept of a multidisciplinary investigation and I determined to follow his example, using a diversity of people from a variety of countries. A multinational team was being employed by Clark and by his French colleagues but their reasons were different from mine. I confess that had I been able to recruit all the scientists I needed in Kenya, I would have done so. As I could not, I decided to ensure that no national group would have special weight or influence on my project.

Kay Behrensmeyer agreed to continue with the study of the geology although her particular interests became quite specialized. She offered to locate and involve other suitable American geologists, who would be needed to map the vast area, and she introduced Carl Vondra from Iowa with this in view. Carl was more than willing and brought in two graduate students who soon embarked upon the task of geological mapping under his supervision. In addition Kay persuaded a well-respected vertebrate palaeontologist, Vince Maglio, to join the team especially to coordinate the collection of fossils and to provide a broad synthesis. He had previously worked in Kenya with the Harvard team on the west side of Lake Turkana. Vince also offered to study the fossil elephants in detail, and advised me which people to invite to study various other fossil groups. Frank Fitch and Jack Miller of FM Consultants agreed to continue their involvement and accepted the suggestion that they should become full-time researchers and carry out the geochronological studies with a research grant rather than on the basis of fees. Fortunately they were able to obtain support in Britain from the National Environmental Research Council and with this support they later introduced graduate students to help in the field work. Two of these students later became principal figures in the project, both carrying out vital geological work that went far beyond the matter of dating.

At the end of the 1969 field season, Glynn Isaac was in Kenya and I decided to ask him whether he would accept the task of coordinating the archaeological studies. I had greatly enjoyed our time together in 1964 at Lake Natron and we were close friends. Glynn agreed to my proposal only after a flying visit to Koobi Fora to see the KBS site and after satisfying himself that my mother was not expecting the job herself. Because of my mother's work at Olduvai she had become a world authority, and had she not had other commitments there she would have been well able to do the work. Indeed, I had asked my mother to describe the first few artefacts in a short paper which was to be published. Glynn, naturally, had been concerned that he should not appear to be taking over her role. Glynn had become Assistant Professor of Archaeology at the University of California, Berkeley, and he envisaged using the project to train graduate students from Berkeley and other institutions. Fortunately he had vehicles and equipment in Kenya from another project and he was able to bring these into the expedition's inventory in 1970. Later, the National Science Foundation in Washington approved a grant towards some of the field costs. We were soon running the project together, Glynn becoming the co-leader.

The biggest question for me was to decide who should study the hominids, especially the two magnificent skulls that we had found in 1969. I was very anxious to have the fossils described promptly but I needed the help of an anatomist. I did not want the material interpreted as to how it related in the overall human evolutionary scheme because I felt that this should best be done later, after a larger sample had been collected. At that time, this approach was rather novel and my father was quick to point out to me that I was breaking with all normal practice. I was soon to cause more consternation because I decided to offer the description of the fossil hominids to Joseph Mungai, a Kenyan, who was then Professor of Anatomy at Nairobi University. Although Joe had never been involved with fossil hominids he was an excellent human anatomist, and agreed to my proposal with the proviso that Dr Alan Walker, another anatomist in his department, should also be involved. At the time Alan, too, had had no direct involvement with the Plio-Pleistocene fossil hominids of Kenya although he had worked on primate fossils from the Kenya and Uganda Miocene during his university training at Cambridge and while working at Makerere University in Uganda. I think my father had hoped to be directly involved himself and when I told him of my plans he was outraged; how could I possibly involve two people who had absolutely no previous experience with fossil hominids? He also disapproved of Alan's involvement because the latter had published a paper challenging some of my father's interpretations.

Alan was pleased to be involved, and looking back now it was one of the best decisions that I made. The other suggestion from Joe was that I, too, should work on the hominid material; he proposed that any ensuing publication would include my name. It was in this way that I first became directly involved with fossil hominids.

My role as leader of the project was to coordinate all the personnel. I also kept up a direct involvement in the search for early man, working with a special team of Kenyans under Kamoya's leadership. This team of young men were amazingly gifted and successful and the large collection of fossil hominids from Turkana is a direct product of their work.

One of my first tasks in preparation for the 1970 field season at Koobi Fora was to build the new camp. Because of the wind it was always very difficult to keep the tents up and sand infiltrated everywhere, which made things even more uncomfortable. The large thatched building with the two Uniports which we had constructed the previous year provided a comfortable dining area and large shady verandahs where we could work in the heat of the day. I now wanted to build thatched buildings for sleeping, permanent toilets, additional stores and a kitchen. This construction began in late May while the main party of scientific personnel came up to Koobi Fora in late June.

The camp we built was delightful; the buildings looked out across the lake which was only about 100 feet away. The lake and countryside constantly change colour, depending upon the position of the sun, the clouds, the wind and the amount of dust in the air. The colours themselves are hard to describe, but in the early morning and late evenings as the sun rises and sets beyond the distant horizons, highlighting the unimaginable vastness of Africa, the colours tend to be pastel – subtle yet dramatic. The lake was cool and despite crocodiles and hippos, everyone swam each day. Bathing in the soft waters of the lake is the most sensuous of experiences. I have not known any visitor to Koobi Fora who failed to be completely captivated by the lake-side camp. I am quite convinced that people perform better when they have pleasant surroundings – guests have been known to refer to the camp at Koobi Fora as the Turkana Hilton! There are ample fish in the lake so that we ate fresh fish regulary. I flew in other fresh food from Nairobi.

An urgent problem that I had to face in 1970 was how to obtain a complete set of large-scale aerial photographs of the area. The year before I had attempted to photograph the area myself in a small plane but the high winds and long traverses made this an impossible task. Stereo pictures had proved invaluable at the Omo to Clark Howell and Yves Coppens for basic mapping and as our area was so much larger it was that much more essential. We knew that detailed geological mapping would take a number of years to complete but we were eager to collect the large quantities of fossils that were lying on the eroded ground. It is essential to relate fossils to the geology of the site where they are found because this provides the vital time reference. There may be other inferences which can be drawn about the climatic or environmental conditions which existed when the animals were alive. In theory we should have refrained from any collecting until the geology was completed, but with photographic cover we could accurately record the position from which any fossil was collected and in this way each find could eventually be tied into the sedimentary sequence when that had been properly worked out.

The photographic project was financed by the William H. Donner Foundation of America in early 1970, but unfortunately bad weather and other problems delayed the completion and we worked the 1970 season with only the poor and incomplete coverage that we had obtained the year before. At the time I was confident about our ability to remember places accurately and I intended that we should mark the spot of each 1970 specimen at whatever time the photos became available. This was foolish of me and it resulted in the precise location of a number of finds being

ABOVE A welcome tea break while *en route* to Turkana

LEFT A young mother with her child in a Mursi village, in the Omo valley, 1967. She has a large clay lip-plug which was a common adornment of the young girls

A few of the family pets. A young hyrax *(top left)* with one of my mother's dalmatians; a tame caracal *(far left)*, kept by one of my brothers; Ben *(left)*, my first dog, who went everywhere with me; our pet otters *(above)*, of course, enjoyed swimming but our dogs were rather unsure

ABOVE A Mursi fisherman at dawn, waiting for the slightest ripple to locate his quarry

BELOW The Mursi men were delighted to have the rather rotten carcass of a dead hippo, which Bob Campbell towed across to them

'lost'. Some magnificent specimens are of greatly reduced scientific value because of my mistake, but I am glad to say that we found only a few hominids, each of which had attracted enough attention to ensure that its position was readily remembered.

By mid-July work had started in earnest. Shortly after the Iowa geologists arrived, the stay of one of them, Gary Johnson, was almost brought to an abrupt end. We had all retired for the night and at about three o'clock in the morning I was awoken by Gary who was calling urgently from just outside my hut. I got up to find him holding his left hand and looking very upset. He thought he had been bitten by a snake. I made a quick examination and decided that he probably had but because nobody had seen the snake I was reluctant to give him immediate treatment with serum. Instead I put a tourniquet on his arm and I cut the finger where he had the bite to induce bleeding. While all this was going on, I sent someone to see if the snake was to be found in Gary's hut. A shout from the hut confirmed the presence of a snake but nobody could tell me what kind it was. I left Gary in the care of a colleague but before I had gone far I had to return because she had begun to feel faint at the sight of Gary's blood! Eventually the snake was killed and I saw that it was a cobra. Poor Gary, his condition worsened visibly with the news!

I managed to keep Gary from seeing the snake and accordingly I was able to represent the offender to him as a 'very small' cobra which would not be at all dangerous. This was not strictly true but I am quite certain that one of the more serious effects of snake bite is psychological. I also gave Gary some serum by injection and I told him how Kamoya had been bitten on the ankle by a much larger cobra the previous year while in bed. Kamoya had not had any serum and he had survived. All this seemed to reassure Gary, and by the following morning he was feeling greatly improved. Unfortunately, it was at this juncture that some well-meaning person found the dead snake and thought that Gary would like to see it. The snake was over four-feet long and my patient had an immediate relapse from which he took a number of days to recover. He was very long suffering and persevered for the duration of the expedition, completing some very useful work.

Kamoya and his team quickly got into the stride of things and soon fossil hominids were being found at a rate of, sometimes, two new specimens a week which, for this kind of fossil, was unprecedented. Meave and I would try to spend each day with the hominid team and I enjoyed some of the most exciting moments of my life in this way. I suppose any hunt is exciting but it is especially so when one is reasonably certain that someone will find something of great interest each day. The team of six would spread out searching over the broken ground and every so often someone would stop, bend down and examine a fragment of bone eroding from the sediments. If it turned out to be something that warranted collection a cairn of stones would be built to mark the spot and we would arrange to return to pick it up later.

Whenever a hominid was found, and this was not every day, the finder would call out to the others and we would all race to the spot. The hominid fossils were more often than not rather poorly preserved but sometimes we were lucky and something really spectacular was found. I always made a point of collecting the hominid specimens myself, being careful to search the site thoroughly for any pieces that might have broken off the fossil.

It is a very strange thing but when one is looking for fossil hominids, one has several subconscious images of the specimens being looked for. I have often thought about this but cannot adequately explain it except in terms of my own personal experiences. To a very large extent, I look for complete skulls and parts of skulls such as teeth, jaws and bone fragments. I can and have walked right past a fossil leg bone fragment, and although I must have seen it visually it never registered as something important. There was a particular case in 1970 that illustrates this very well. I had been away from Koobi Fora for a day to visit Kitale where I had to buy supplies and it was during this time that Meave had found the first fossil hominid femur (thigh bone) ever to have been recovered in Kenya, let alone at Turkana.

The bone lacked both ends and looked sufficiently different from the same bone in a modern human that she was unsure what it was although she suspected it might be hominid. Before 1970, fossil human limb bones were very rare in the collection even though greatly prized as useful scientific specimens. Skulls, jaws and teeth are much easier to recognize and they give a great deal of interesting information about the creature's diet and possible state of brain development. Limb bones, however, are also of considerable value because they tell us about posture and how our ancestors moved. The complex process of developing bipedalism and the freeing of the hands is of vital interest to our own evolutionary story. One femur had been found at Olduvai by my mother and her team the year before, but that was virtually the extent of the finds in East Africa. Meave's find caused great excitement at Koobi Fora and considerable discussion as to whether it really was hominid. When I got back from Kitale and it was brought out for me to examine, we sat around the table talking about it. It was just after a late lunch, and as I admired Meave's 'leg', there was a stirring in my subconscious; I began to remember seeing a bone rather like it myself. The problem was that I must have seen it more than a year before. Was it the same bone that I had not recognized or could there be two?

In the afternoon, Meave took me to the place where she had found the femur and I became even more sure that I had seen a second very similar bone not far away and I determined to look quietly for it on my own. The next day I spent the morning in camp writing and at lunch time I told those in camp that I was going to fetch the men who had been sieving a hominid site not far from where Meave had found the femur. I slipped off to search for my bone.

I began to retrace my steps in the area that I had explored all those months before and I became more and more excited because, as I progressed, every detail seemed so clear in my mind. I had been here before: I had seen that pig jaw, I had examined that hippo tooth, I had been scratched by that branch, I had slipped on the scree here and, if I was right, my bone should be over that knoll. There it was!

Meave records in her diary that 'The others returned at about 1.45 p.m. but Richard was not back until 2.0 p.m. I wondered why he should be so late and why he looked so pleased with himself but he merely explained that he had had a puncture. After lunch when everyone had gone he excitedly said to me "Come and see my little bone!" He had found a femur very like mine but more slender and graceful and more of it. I was amazed and wondered how he did it but he simply remarked that he had to maintain his position!' Since then a large number of limb bones have been found because we all began to look subconsciously for other parts of the skeleton and not just for skulls and teeth.

A corollary to this story occurred years later. I had spent a two-week holiday with my family and friends at Lamu on the beautiful Kenyan coast north of Mombasa. One of our favourite sports there was to scour the reefs for crayfish or tropical lobsters. These tasty beasts rest under overhangs of coral and you can often spot them by the tell-tale sign of their waving feelers. To catch one you dive down, grab the feelers and pull the kicking lobster from its place of concealment. All of this is easier to write about than to do, but there is a lot of pleasure to be had from simply searching for the lobsters in the first place. In any event, shortly after this holiday I returned to Koobi Fora to look for Turkana fossils. After a full day in which I failed to find any fossils of interest, I realized that my subconscious was making me look for lobsters! The next day I made it very clear to myself that the image in my mind had to be of bones and by this conscious effort I was soon seeing interesting fossils again. Maybe this also explains why I so seldom see lobsters when I am out on the reef!

The joy of searching for fossils in remote and difficult places is that there is always a strong possibility that each 'find' will tell you something new. At Turkana fossils are very common indeed and you can hardly walk ten paces without seeing a fossil of some kind. Many are simply broken scraps of bone that cannot be identified, but every specimen that is collected adds to the sum of the knowledge about a particular species and its former existence.

For example, at one time, it was thought that camels had been introduced into Africa as domesticated animals relatively recently. This idea was successfully challenged when some fossil camel foot bones were found at Olduvai. Later, others were found at the Omo but, nevertheless, all that could be said with certainty was that camels were undoubtedly part of the African fauna two million years ago. This much was known before we went to Lake Turkana, but I kept a mental note in the back of my mind to look out for fossil camel bones. One day, one of Kamoya's team found part of a camel's lower jaw with teeth in deposits that were about three million years old. Seeing this specimen for the first time lying exposed on the sediment I realized at once that this one specimen would provide exciting evidence of how long camels had been in Africa. In due course other finds were made, and an important scientific paper has been prepared by John Harris, on the early camels of Africa. He is a palaeontologist, who worked for me at the National Museum between 1970–79, and he is particularly knowledgeable on fossil pigs, antelopes and other large African mammals. He subsequently became Meave's brother-in-law.

We had riding camels with us again in 1970, but by this stage we had got to know the area sufficiently well for it to be an easy matter to get vehicles to all the fossil sites. So, although some of our happiest days were when we were with the camels, our primary purpose was research and we no longer really needed to use them. These animals are particularly useful when distances are too great for walking and when vehicles might crush fossils on the surface. On a previous occasion, for instance, I had visited a site on the western shore of Lake Turkana where I was horrified to find vehicle tracks right over fossils that were partly exposed.

It is so important to be able to travel speedily and efficiently when organizing the sort of large-scale operation that I was, that I decided I must take up flying again. Although I had acquired my pilot's licence in 1963, I gave up piloting in March 1968 following a stupid incident at

Olduvai the previous year in which I very nearly killed several good friends. I was trying to land against the early morning sun which was so bright that I was quite unable to see the markers on the little grass airstrip that I had built myself some weeks before. I made three attempts but each time I aborted the landing at the last moment when I realized that I was not straight on the runway. I became very anxious; my pride was at stake. Instead of going away or attempting the landing from the other direction I foolishly determined, on the fourth attempt, to put the plane down regardless. This I did but hit the rocks marking the edge of the runway and in a few moments we had shed two wheels, the tailplane and half of one wing! Fortunately I remained calm and managed to turn off the electrical switches in good time. Luckily there was no fire and none of us was physically hurt but I was thoroughly ashamed and very shaken.

It took some effort to learn to fly again because I had to regain my nerve but after a few difficult flights in early 1970, I soon got back to a point where I was reasonably safe again and I regained my licence. I hired a small four-seater Cessna for the duration of the 1970 expedition and it made a tremendous difference. The regular flights soon dispelled my fear and although I still do not enjoy flying, it is not a traumatic experience every time I do!

The flight from Nairobi to Koobi Fora is fairly straightforward except for the odd occasions when bad weather makes detours and low flying necessary. Our landing strip near the camp is unpopular with pilots because it is very short; a total runway length of 428 yards and often there is a strong cross wind that can reach 25 knots at a 75° angle! I have now landed and taken off from this field well over a thousand times without incident but it always calls for special concentration and effort.

Many friends have asked why we do not extend the length of the strip and the answer is very simple. If we did all sorts of uninvited people would fly in to Koobi Fora and it would become a real nuisance. By having an airfield that is tricky to use visitors are forced to use the larger strip which is about six miles away from our camp. To visit us they require ground transport and so their visit has to be arranged in advance with our consent. In this way, we have a very effective visitor filter.

In fact, I have numerous little airstrips scattered across the whole area. Just as I abandoned the search for fossils by camel in favour of Land-Rovers, I eventually gave them up in favour of an aeroplane. We were operating several camps during the 1970 season because some of the sites were too far from Koobi Fora for daily commuting. Glynn had a large camp consisting of tents for about thirty people, close to the KBS site where he was excavating. Kamoya had a smaller mobile camp that moved regularly depending upon the area in which he was working. We had also established a camp to the north near Ileret where Vince Maglio spent several weeks collecting fossils.

I spent an increasing amount of time on logistics. It is no easy task to keep a group of up to 70 people adequately provisioned when the closest source of supplies is over two days' drive away and furthermore I had to keep in touch with all the differing activities of my multi-disciplinary team. The disadvantage was that I actually did very little field work myself. If at all possible, I would spend a week at the lake followed by a week on Museum business in Nairobi.

The value of our small aeroplane was so clear that I made up my mind to seek funds to purchase our own plane instead of renting one. I was fortunate; with a grant given to me jointly by the National Geographic Society and the William H. Donner Foundation, I was able to buy a second-hand Cessna Skywagon in September 1970. The mobility this gave enabled me to provide the leadership such an ambitious and diffuse operation demanded.

It is very difficult to assess the success of any particular year, but there is no doubt that the Koobi Fora-based project really got under way in 1970. As the weeks went by our highest hopes were surpassed by Kamoya's continual discoveries of fossil hominid remains. The sheer joy of being successful was very invigorating and morale among the group was very high. It was abundantly clear that the Turkana site was amazingly rich in well-preserved fossil hominids and I was convinced that many questions on human origins would be answered as a result of our work.

It was in October 1970 that Meave and I got married. The District Commissioner in Nairobi, Bill Martin, who was also a very dear friend, married us in his office. Bill is Kenyan and a long-standing Trustee of the National Museum; he had, in fact, also married Margaret and me.

When it became clear that we would be spending a lot of time at Koobi Fora in the future, Meave and I decided to build ourselves more comfortable accommodation. I chose a site on top of the ridge, set away from the other camp buildings so that we could be private and a little separate from the expedition. It was a beautiful situation overlooking the bay and with a cool wind blowing off the lake. The house that I built had a stone floor, thatched roof and thatched walls, although the insides of the walls were lined with timber. It had a main bedroom and a smaller second bedroom which we used for a study. Between the two bedrooms was a small store where we kept a kerosene-fired refrigerator and we had some shelves for the storage of our own personal effects. Over the years we accumulated a considerable quantity – nothing important but things to which, inevitably, we had grown attached.

The loss of this house three years after it was built was one of my great tragedies at Koobi Fora. We were in Nairobi when this occurred; I simply received a radio message that our house had burnt down. Meave and I flew up immediately to find all that remained was a pile of ash and charred rock. One of the camp staff had been filling the refrigerator burner with kerosene and after relighting it, a small fire developed where some of the fuel had spilled on the tank. Realizing that a dangerous situation was developing he moved the tank from underneath the fridge and carried it, burning furiously, out of the house. He thought that the best place to put it would be in the lee of the house away from the wind where he hoped to extinguish the fire with sand. Unfortunately, the tank exploded and the flames were carried into the dry thatch. Luckily, he was not badly hurt but within moments the whole house was alight.

Gallant attempts by Ian Findlater, an English geology student, and some members of our staff to rescue our belongings were terminated when ammunition started exploding in the building and everyone had to take cover. The heat must have been tremendous. We found the remains of glass bottles that had melted and one even had a bullet embedded in it! One of the most gruesome reminders of the completeness of the fire was the charred remains of several human bones, laboratory specimens of recent age that we had in the store for comparison with our fossil finds. Looking at them lying in the dark ash made us deeply grateful that no one had died in the fire. We subsequently rebuilt the house on the same site but in stone with only the roof thatched.

From 1969 onwards I was able to operate at Koobi Fora without an armed police escort but security remained a natural concern. Since the camp had been vandalized I was well aware that there were bands of people who came through the area and, as I have mentioned, we knew from police reports that these men were often well armed. I had my own rifle and Kamoya was similarly armed but what could the two of us do if we were attacked? My hope was that the situation would never arise; I trusted that any scouting party would report that we were simply not worth the trouble of raiding. In spite of this, I did urge everyone to be on the look out. The first real excitement came quite unexpectedly one evening in early September 1971.

We had finished dinner and it was about ten o'clock at night. The moon was very bright, reflecting off the dappled lake surface, and the sky above was clear, alive with brilliant stars that are so typical of the African desert sky. As I walked up the path to my house, I turned to survey the camp. Everyone was retiring for the night, the oil lamps were being put out and a sense of tranquillity hung over the camp. Then I saw what looked like a canoe passing across the moon's reflection in the lake; for a moment there was a clear silhouette, but then it was gone. I waited and again I saw it – this time I was in no doubt. I quickly fetched my binoculars and sure enough, I could count a total of thirteen canoes, each carrying about six men and silently moving as a flotilla near the shore. Who were they and what was their purpose? I had no idea but I realized that if they were to attack us, the sooner we evacuated the camp the better.

I rushed back to the other buildings and quietly raised the alarm; everyone was to assemble at the vehicles at once and complete silence was to be observed. In a few moments, my entire expedition was gathered and at least twenty people had jammed themselves into one Land-Rover, anxious to be the first to be evacuated! I asked for five volunteers to stay in camp with me as a 'home defence'; we had one rifle and some firecrackers and I intended to post sentries, just in case. Glynn took the rest of the party inland to spend the night hidden, well away from the lake shore.

Just before Glynn moved off with the evacuees, one of my 'volunteers' pleaded to be allowed to join Glynn on the grounds that he was a pacifist and had refused to fight in the Vietnam war. I was furious but let him go.

The night was sleepless for everyone but quite uneventful. The canoes were soon lost from sight as clouds shielded the moonlight. We waited in suspense, anxious for the dawn. The next morning there was no sign of the canoes so I took off in my plane to see if I could discover where they had gone. After some searching, I spotted them concealed in the reeds some two miles south of our camp. Careful observation showed about fifty men, some with rifles, but it was clear that their destination was to the south and we were not of any interest.

On another occasion, our truck was travelling from Koobi Fora to Nderati to fetch water when it came over a ridge and surprised a party of about seventy men who were resting in the shade of some trees. Possibly out of fear or as a warning they opened fire on the truck, but by good fortune the range was too great and there were no casualties. Needless to say, our truck driver refused to go for water again unless accompanied by an armed escort. We learned later that this gang had gone on to raid a village about seventy miles away and a number of people had been killed.

Our ferry, with Land-Rover aboard, crossing the Omo without incident

ABOVE Crocodiles, some very large, were common all along the Omo River

LEFT Speeding up the Omo in our aluminium boat, with my father and Margaret

Until 1976 we had had no domestic animals in camp – the area was wild and the idea of keeping such creatures had never entered my head. One day I flew into camp and I was surprised to notice that there was a dog hanging about near the kitchen. I asked for an explanation of this quite unexpected sight. I was told that some days before my arrival this dog had wandered into camp looking extremely thin and dejected and that it had been scavenging from the rubbish pit at night. I was also told that some people in the camp were frightened that the dog might be rabid and that they had made every attempt to drive it away by throwing stones at it. I suggested that the best thing to do would be to offer the dog some fresh meat and a bowl of milk. The dog was adopted as a full member of the camp and named Nzuma, the wanderer. Nzuma, it appeared, had been with a party of nomads who were ambushed by another raiding party at the water hole at Nderati. Seven people had died and their dog must have wandered for several weeks before he found our camp.

Nzuma never ceases to amaze us. For a number of years, until we acquired a second dog, he lived alone at Koobi Fora and in order to get company he befriended the local jackals. He would frequently go off with the jackals for four or five days, travelling thirty or forty miles from our camp, and in those circumstances Nzuma lived as a wild animal feeding with the jackals or hunting for himself. It is not uncommon for him to bring back to camp the remains of hares or even young gazelles. Interestingly, he has also successfully bred with the jackals. Extraordinary as this is, perhaps even more curious is the fact that the progeny have been fertile and they too have bred. At Koobi Fora today there are at least two animals that are half dog and half jackal, to my knowledge a unique example of crossbreeding between a domestic dog and its wild counterpart.

Despite the trouble we take to be methodical, sometimes the most exciting finds are still pure accidents. One morning in 1971, for instance, three of Kamoya's team were sitting round a sieve, quietly looking for tiny fragments of bone from a hominid lower jaw that had been found several days before. Apart from the low murmur of their voices there was no other sound of human activity. It was a particularly still and peaceful scene with the gentle rustle of the wind in the bushes and the periodic call of the African Laughing Dove contributing to the mood of tranquillity. Suddenly a shot rang out and there was the unmistakable sound of a bullet hitting the hard ground very close at hand. The three men at the sieve fled for cover as fast as they could, running one hundred yards before they reached a bush high up on a river bank. They hid there for at least an hour and fortunately nothing else happened and they never saw a soul. As they came out from under the bushes one of them almost trod on a complete fossil jaw of a hominid partially exposed on the bank. Had they not scrambled up the bank to hide after being shot at, that specimen would probably never have been found and one more piece in the jigsaw of man's origins would have eluded us.

CHAPTER ELEVEN

THE DISCOVERY OF '1470'

THE YEAR 1972 WAS AN EVENTFUL ONE. Meave gave birth to our first daughter, Louise, on 21 March; the best known of all our discoveries from Kooba Fora, the skull '1470', was made in July; and in October my father died. After the birth of Louise we had slightly to rearrange our lives. Fortunately we both agreed that children should not be pampered and from the age of six weeks, Louise came along on all our trips. I am quite certain that Louise and her young sister, Samira, who was born two years later will look back with pleasure on their trips with us as young children. It is sad that my first daughter Anna did not have the same opportunity as them. Koobi Fora is very hot as well as dry and the greatest difficulty was seeing to it that Louise did not become dehydrated. Meave made a netting bed that could be hung from the rafters of the house and, provided there was a breeze, Louise never got as hot as we did!

The skull '1470', the earliest evidence we had for *Homo* at Koobi Fora, was discovered by Bernard Ngeneo who, although he had only joined Kamoya's search team the previous year, quickly became an accomplished fossil hunter. Bernard came to be a member of the expedition in a rather unusual manner. I have several domestic staff at my home just outside Nairobi but early in 1970, when I returned with Meave from a trip abroad, I discovered Bernard Ngeneo 'helping out' with the housework. I assumed he was a friend of my regular helper and as everyone seemed content I did not say anything, fully expecting the arrangement to end in a few days. After a month, Bernard was still there and he had been so helpful that I decided I must pay him for his work. But I also determined to find out about his longer-term intentions and I explained that I was not in a position to offer him a job. Bernard was quite nonplussed by this and asked if he might go on helping his friend for the time being. This state of affairs continued for several months and in due course the Koobi Fora season began and Meave and I prepared to leave.

As we drove out of our driveway we found Bernard standing by the road with his small suitcase. I offered him a ride thinking that he was on his way into the city and it was therefore easy to drop him off at the airfield from where he could catch a bus into town. While driving, I casually asked him where he was going and he equally casually replied Koobi Fora. I was so impressed by his enterprise that I decided to take him with us and gave him a temporary job helping in the camp kitchen.

So began Bernard's long association with the work at Turkana. At the end of August he asked if he might see something of the area where the scientists were working and I arranged for him to accompany Kay and Glynn on a day trip to the KBS site where there were geological samples to be carried and various other chores that he could usefully do. While Kay and Glynn were talking at the site, Bernard wandered off and found an australopithecine thigh bone which was lying only 100 yards from where Glynn and his team had been working for almost two months! Needless to say this earned Bernard a place on the expedition's prospecting team. Like all of the other skilled fossil hunters, he learned the details of skeletal anatomy by simply handling animal skeletons and bone fragments; comparing one species with another. This was done in the field as well as in the Osteology Laboratory at the Museum. During the subsequent years he found a number of important specimens.

The 1972 discovery of '1470' has had tremendous publicity and is certainly the best-known fossil from Koobi Fora. When found, however, it caused no real excitement other than the usual good feeling that another hominid had been discovered. I was away in Nairobi at the time, but when I visited the site several days later on 27 July, everything was just as Bernard had found it, nothing had been disturbed. The specimen was badly broken and many fragments of light-coloured fossil bone were lying on the surface of a steep-sided ravine. None of the fragments was more than an inch long, but some were readily recognizable as being part of a hominid cranium. One good thing that was immediately apparent was that some were obviously from the back of the skull, others from the top, some from the sides, and there were even pieces of the very fragile facial bones. This indicated that there was a chance that we might eventually find enough pieces to reconstruct a fairly complete skull. It was clear, however, that a major sieving operation was required to recover other fragments that might be lying buried in the top few inches of soil or which had been washed down the steep slope. This sieving operation was not begun until a fortnight later and it continued over many weeks.

A number of fragments were collected in the first few days of sieving. On the fifth day, Meave, Bernard Wood (a friend who had been with me on several previous expeditions) and I flew to the site to help. At lunch time we returned to Koobi Fora with a number of fragments and after eating and a welcome swim we retired to the shady verandah of our house to examine the pieces. Meave carefully washed the fragments and laid them out on a wooden tray to dry in the sun and before long we were ready to begin to find which pieces could be joined to others. In no time at all, several of the bigger pieces fitted together and we realized that the fossil skull had been large, certainly larger than the small-brained *Australopithecus* such as we had found in 1969 and 1970. By the end of that exciting afternoon, we knew that we could go no further with the reconstruction without more pieces from the sieving.

Over the next few weeks more and more pieces were found in the sieving and Meave slowly put the fragments together. Gradually a skull began to take shape and we began to get a rough idea of its size. It was larger than any of the early fossil hominids that I had seen but the question was, how large was the brain? We decided to attempt a crude guess. Beginning by carefully filling the gaps in the vault with Plasticine and sticky tape, we then filled the vault with beach sand and measured the volume of sand in a rain gauge. By a most complicated conversion we came up with a volume of just under 800 cubic centimetres. The actual value for the brain size of '1470' has since been established by accurate methods as 775 cubic centimetres, so we were very close. This was fantastic new information. We now had an early fossil human skull with a brain size considerably larger than anything that had been found before of similar antiquity. Also, we had found some limb bones. At the time we believed that the skull must be older than 2.6 million years – this being based upon the dating of the KBS tuff and the assurances that we had from John Miller and Frank Fitch to the effect that this was a good date. It turned out that we were wrong by at least half a million years but this we only learned much later.

Not long after, I had to return to Nairobi. I took the skull with me because I was really anxious to show it to my father who I knew was planning to travel abroad in late September. I actually showed it to him on the morning before his departure for England. We had a long and extraordinary discussion in his office after which it was decided that I should join my parents for dinner to continue talking about the new find. Meave was still at Koobi Fora and I flew back the following day to collect her and our daughter Louise.

The meeting and discussions with father that day remain clear in my memory because, for the first time in many years, he was completely relaxed and at ease with me. He gave me the impression that he was really pleased and delighted by our find of '1470' and there was no tension between us. Seeing and handling the '1470' skull was an emotional moment. It represented to him the final proof of the ideas that he had held thoughout his career about the great antiquity of quite advanced hominid forms. No discovery of his had been as complete and he felt sure that there would no longer be serious doubts about this. He was delighted that it was a member of my team who had made the find at a Kenyan site. In many ways I felt closer to him that day than I had since my early childhood. For reasons I cannot explain, I had a distinct feeling that evening that we had finally made a real peace and I was aware that something quite extraordinary was happening between us. Curiously I had a nagging worry about his impending trip and wished then that we had more time to talk.

At the airport late that night, I bade him farewell with the strongest sense of parting. In fact we discussed it just before he went through the departure gate and I urged him to take particular care of his health. That was on the night of 29 September; he died in London after a severe coronary on 1 October. It is hard to believe but it was as though we had both had a premonition.

The huge number of sympathetic messages my mother received from all over the world made a great impression on us. The Government of Kenya went out of its way to help the family and we received personal messages of condolence from the Head of State and many other national figures. We were comforted to feel that my father's long service to our country was recognized and that many Kenyans felt my father's death to be a loss that Kenya herself had suffered.

One of the things that father had asked me about was my plans for releasing the news of the discovery of '1470'. It was his view that I would receive considerable opposition from some colleagues who would find it impossible to accept the new skull as *Homo*. It so happened that I had been invited by Dr H. G. Vevers, then Assistant Director of Science of the Zoological Society of London, to present a paper during a symposium the society was to hold in November. A number of other people also involved in the study of fossil man in Africa, were invited to present papers as well. This particular symposium, the brainchild of Dr Vevers, was organized jointly by the Zoological Society and the Anatomical Society to commemorate the birth of Sir Grafton Elliott Smith. Born in Australia in 1871, the latter became a leading British anatomist, and was involved with the Piltdown skull. The Zoological Society is a prestigious organization that has a respected publication record and so it seemed an ideal venue for presenting my recent discoveries at Lake Turkana. I was, in fact, particularly anxious that my new find should be presented to the scientific community so that I could obtain its reactions before the popular press made any announcements. As it turned out, the scientific gathering received the details in the morning, and in the afternoon I gave the basic facts about the new skull to the newspapers at a specially arranged press conference.

I wrote to Dr Vevers telling him that I wished to speak about my exciting new discovery and that my friend Bernard Wood would call on him to discuss the question of science reporters. I received a reply from Lord Zuckerman, then Secretary of the Zoological Society, who pointed out that a meeting organized to commemorate the memory of Elliott Smith was hardly the occasion for me to have a press conference, that such an episode would detract from the scientific value of the occasion and of my work, and he hoped that I would agree. Although I truly believed that my new skull would create so much interest that the Society must benefit from the announcement and that its public image would be enhanced by holding a press conference later in the day, I was quite prepared to go along with the decision.

Several weeks later I arrived in London and at 9.30 a.m. on 9 November I presented myself at the meeting rooms of the Zoological Society of London at the start of the day's sessions. The anatomist, Professor A.J. Cave was in the Chair and after some introductory remarks about the symposium, Lord Zuckerman was invited to give his lecture on Sir Grafton Elliott Smith during which he touched on the Piltdown skull and the fact that Smith could certainly never have been one of the hoaxers. The Chairman then called upon Professor J. S. Weiner to give us more details of the Piltdown affair.

There was just time for one of the other three scheduled speakers to present his paper before the coffee break and by this time the Chairman became very strict about keeping time. He told us that as the meeting was running late, the time for each speaker had to be reduced. During the break everyone gathered in groups to discuss the morning's proceedings.

I had thought that I would have 50 minutes to speak and found it extremely difficult to decide at the last moment what to delete from my carefully prepared paper, in order to reduce it to the new time of 30 minutes. I decided that the best thing to do was to abandon the original text completely and speak *ad lib*. I spoke briefly about the Koobi Fora site, making reference to the biological framework to which various fossils could be related. I spoke of the dating work which had been carried out by Fitch and Miller, who were in the audience, and I expressed some confidence that the hominid discoveries from this area would prove crucial to our understanding of human evolution. I then made reference to the various specimens of *Australopithecus* that had been recovered and remarked upon the apparent similarity between the material from Koobi Fora and that from Olduvai.

At this juncture I introduced the new skull and, using slides, I carefully described the salient features, stressing the large brain size and remarking that in my opinion this specimen could not be assigned to the genus *Australopithecus* but should be attributed to *Homo*. I concluded my presentation by summarizing the overall picture that was emerging from the work at Koobi Fora and drawing attention to the fact that the new *Homo* skull and limb bones were, as I then believed, at least 2.6 million years old. I drew attention to the fact that *Australopithecus* could now be seen to have been a contemporary of *Homo* for a long period and that the previously assumed relationship, in which *Australopithecus* was thought to be an ancestor of *Homo*, should be reconsidered. I thanked the Chairman and sat down to heartening applause led by Lord Zuckerman.

As things quietened down, Lord Zuckerman rose to speak wishing, as he put it, to be the first to congratulate me on my presentation. I quote directly from the *Proceedings*:

'Mr Chairman, may I first congratulate Mr Leakey, an amateur and not a specialist, for the very modest and moderated way he has given his presentation. May I also express my personal gratitude, and certainly the gratitude of many others who have worked with him and his father, for the work they have done, not as anatomists, as Mr Leakey pointed out, not as geochemists or anything else, but just as people interested in collecting fossils on which specialists work. The generous offer that these are available to students is certainly something that we shall take great account of, and I trust use.'

Several other members of the audience asked relevant questions and after about fifteen minutes, the Chairman adjourned the meeting for lunch. All the participants were to have a private lunch in the Fellows' dining room across the road. I happened to walk out with Lord Zuckerman ahead of all the others and as we came into the foyer, there must have been twenty or more reporters, cameramen and others. I was as surprised as Zuckerman seemed to be furious, especially when he realized that the correspondents wanted to talk to me about the new skull. I was quite innocent, for I had not told anyone in the Press that I was attending the Zoological Society meeting. Lord Zuckerman led me rapidly into an inner office where I was asked to wait. Not long after, a staff member of the Society appeared and asked me what sort of sandwiches I would like to have for lunch! I explained to him that I intended having lunch with my colleagues but I was informed that I was to remain out of sight until the Press had left the premises. As the reporters seemed inclined to wait, lunch was brought in to me. It appeared to me that I was being held, graciously I grant you, against my will.

After a while I managed to contact my colleague Bernard Wood, and I persuaded him to ask the reporters, who were still waiting, to meet me somewhere after lunch. I suggested the venue be Kenya's High Commission and after brief discussions with the officials there, all was arranged and the correspondents left the premises of the Zoological Society. I was then free to join the others.

The press conference turned out to be a great success and the worldwide coverage was extraordinarily thorough. I can only imagine that the world just happened to be quiet that day because it is unthinkable that under normal circumstances a fossil skull would make the front page in so many papers all over the world. Of course publicity breeds publicity. Feature editors were impressed by the importance that their news editors had given the skull and, without really knowing why, they too gave the skull VIP treatment.

I have never really made up my mind whether the publicity was a good thing or not. It certainly gave an exaggerated importance to the individual skull known as '1470' and many people seem to believe, quite wrongly, that only one really important find has been made at Koobi Fora. In fact, '1470' is just one specimen in a couple of hundred and there are at least half a dozen others that are just as complete and just as important. Nor was the skull the first *Homo habilis* to have been found. Indeed, in 1962 my father had discovered a skull of *H. habilis* at Olduvai which received virtually no publicity.

The whole question of whether a skull should be called *Homo* or something else is a matter of definition. None of the fossils that we find are labelled. We give them names for our own convenience. We have to judge whether 'X' looks more like 'Y' than 'Z' and this decision is often made more difficult because 'X', 'Y' and 'Z' are incomplete. I called this particular skull *Homo* because I believed it to be more like other fossils that had been called *Homo* than it was to those called *Australopithecus*. More importantly, '1470' has a brain size which is considerably bigger than any of the known fossils of *Australopithecus* and this is, in my opinion, very significant.

The intelligence we have, along with our technology and culture, all stems from some event way back in time when it was advantageous to be larger brained. My interest in early *Homo* is nothing more than a desire to determine exactly when the brain began to increase in size and there is no doubt, even after the revision of the dating, that '1470' is one of the earliest examples of a large-brained hominid. What I also want to know is whether the brain enlargement that occurred actually happened to a species of *Australopithecus*, or to a quite different hominid.

The announcement of '1470' in London in 1972 began this debate as far as I have been personally concerned. There were many questions that had to be dealt with; one of these was whether or not we had correctly reconstructed the skull from the many fragments. It was suggested that the new skull from Koobi Fora was incorrectly assembled and if it were redone correctly, the skull would be typical of *Australopithecus*. The fact of the matter was that we were able to recover a sufficient number of skull fragments to provide a continuous surface from the front of the skull to the back and from side to side. All the pieces fitted together perfectly and there were no floating fragments whose position was uncertain. Under these circumstances, the reconstruction of the skull has to be correct. There is some deformation of the skull as in most fossils, caused by pressure and movement of the rocks in which it was buried. The real size of the brain-case, however, cannot be greatly altered by correcting for this deformation. Unfortunately so much time and energy has been spent on trivial arguments over the '1470' specimen that its real importance is often overlooked.

Since the discovery of '1470' in 1972, other less complete skulls of the same type have been found and these have confirmed that the specimen is not unique. The more complete specimens from Koobi Fora have helped to clear the controversy and most scientists now accept that there is a larger-brained hominid, distinct but contemporary with *Australopithecus*, at Olduvai and Koobi Fora. Most people would now agree that '1470' should be called *Homo habilis* and that it is a direct ancestor of *H. erectus*.

When father died, I suggested to the Museum Board of Trustees that it would be appropriate to construct a modern building for housing the collections of archaeological and palaeontological material stored in various corners of the Museum. I envisaged a building large enough to allow for a considerable expansion of the collections as a lasting memorial to my father. It seemed an ideal opportunity to get a proper vault for the valuable hominid fossils as well as other badly needed facilities. I was very confident that I could raise the money for the memorial and I saw this new building as a base for an expanded international programme in prehistory.

I had first floated the idea of an expanded facility in 1971, but for some reason father was opposed to it. His main objection was that we did not need 'fancy' buildings and that the money raised would be better spent on paying salaries. For various reasons I chose not to fight this initial response, and the idea was shelved. However, after his death, it seemed an appropriate way to commemorate and continue the work of a pioneer prehistorian and so it was agreed that, provided I raised the capital, the National Museum would construct a new set of buildings to house the prehistory collections. The buildings would also incorporate new administrative offices for the Museum, a board room and a 500-seat lecture hall for the general use of the Museum.

I began the task of fund raising in February 1973. One of my first objectives was to seek the support of a suitable foundation in the U.S.A. Many people in America will generously support projects provided their money can be given to an organization that qualifies for tax deduction. This is because they are then only taxed upon the income remaining after such deductions have been made. There is no doubt that these deductions have enabled many scientific and educational projects in America to get off the ground and it is unfortunate that more countries have not provided businessmen with the same incentive to support worthwhile projects. The system was undoubtedly abused by some and in 1969 the American Government passed stringent legislation to ensure that foundations attracting money for charitable purposes were not only genuine but were also broadly based so that funds could be donated to a number of enterprises. As a result it became more difficult to obtain foundation support for the sort of fund raising I had in mind. No foundation could afford to act simply as a channel for funds because the 'conduit' could lose its tax deductible status if the authorities of the American Internal Revenue Service so decided. What I needed was a foundation that had as its main charter the support of research in African prehistory. As it happened, such a foundation actually existed. The LSB Leakey Foundation, which my father had spent so much time and effort supporting, had for a number of years provided generous financial assistance to several of his projects. I had given lectures for the Foundation from time to time myself and in 1973 I knew some people on the Board. I had also been somewhat critical of the way in which the Foundation functioned and I had made my feelings known. Nevertheless, during father's lifetime, I kept myself as distant as I could from its affairs, not wishing to be in unnecessary conflict with him.

After the death of my father I decided to change this and on a trip to the

United States in early 1974, I held discussions with members of the Board. I believed that I could help the Foundation in my efforts to raise money for my father's memorial and that they could help me. My idea was that I would spearhead the fund raising and that the money produced by me would be channelled through the Foundation. I had to raise more than a million dollars for the memorial building and I needed to be able to work through an organization such as the LSB Leakey Foundation. For my part, I wanted the Foundation to become more concerned with prehistory and less involved with studies in primate behaviour and other more general projects. My ideas were presented in a memorandum which went before a special meeting of the Leakey Foundation Board. Among my suggestions was one for a new board which, on reflection, was impertinent and unreasonable, although at the time it seemed entirely justified. In spite of this I was confident of success because a majority of those present at the meeting had spoken to me individually indicating that they would back my ideas.

At the meeting I was asked a number of questions about my memorandum, and then excused. About an hour later I was recalled and told that my proposal had been rejected and that I could not reasonably expect the Foundation to accede to my suggestions. At the time I was surprised and before leaving I decided to comment. I expressed my regrets at their decision and informed the meeting that I would have no alternative but to set up a new foundation that would have the more specific focus that I wanted. I thought then that this would be unfortunate but that in the final analysis it was essential for there to be an American foundation which would have a special relationship with the proposed building in Nairobi. I concluded by proposing that Board members might need more time; I sought a final written decision within ninety days and on this basis I agreed to delay setting up a new foundation.

It was during this visit to the U.S.A. that I persuaded my friend David Look to help me raise money for the memorial building. With his aid in talking to foundations, corporations and wealthy individuals the required $1.5 million was made available. No friend has ever done more for me and I shall always remain in debt to the Look family. What is amusing is how we met in the first place.

When Meave was a student in England, she spent a summer vacation in America, during which she lived for a couple of months with the Looks in New Jersey caring for their four children, and so earning some pocket money for her trip. She had kept in touch with the family since then and on a number of occasions David, who was involved in banking, came to

Nairobi on business and would contact Meave. I never met him in Nairobi – I think I resented the idea that my wife had been *au pair* to David's family. That this was illogical and quite silly did not occur to me. In any event, in 1971, Meave was with me in America and I had a lecture in New Jersey. Meave decided to visit the Looks and we agreed to go our separate ways. I was collected by a large, luxurious limousine, while Meave went by Greyhound bus! It turned out that we were in fact going to almost the same place, no more than six miles distant. The Looks were very interested in our work at Koobi Fora and when Meave explained the problems of raising money for the research they offered to help. Later on in the trip, I went with Meave to the Looks for lunch on a Sunday and David and I became firm friends.

By the end of March 1974 it seemed that the Leakey Foundation was not going to respond and I began to plan the setting up of a new foundation. I had in fact already spent some time discussing things with David Look and another friend, Charlie Jaffin, who was a New York attorney. I was confident that we could move quickly if the need arose. I was about to contact David and Charlie in America when I received a cable to the effect that the President of the Leakey Foundation was arrivng the next day in Nairobi for discussions on my proposals of some weeks earlier. This was unexpected and I awaited his arrival with considerable interest and impatience.

It turned out, as far as I could determine, that the Leakey Foundation had reconsidered my ideas and while unable to accept them as a package, a counterproposal had been worked out for my consideration. This fell far short of what I wanted and I could see no advantages in taking it up. I expressed my appreciation for the Leakey Foundation's effort to meet my wishes and I informed my visitor that I would be going ahead with plans to establish an alternative foundation which was to be called the Foundation for Research into the Origin of Man (FROM for short). I cabled Charlie Jaffin and within three weeks the papers had been filed and FROM was in business, so to speak. We were given a two-year public foundation and at the end of this term we would be subject to audit and a ruling by the Internal Revenue Service.

David Look became President, Charlie Jaffin agreed to be the Secretary and I was elected Chairman. A number of friends consented to serve on FROM's first board of directors and these included two scientific colleagues, Glynn Isaac and Don Johanson. David undertook to run the Foundation from his desk in the Morgan Bank for a while to enable the whole thing to get off to a good start without the burden of an office and all the associated administration. At this early stage I had not really come to terms with the fact that FROM could not be used specifically to channel money to the Museum in Nairobi because of the tax laws in America. However, my legal friends wasted no time in putting me straight on this point and I had to find an alternative. I needed a charitable organization to take on the funding of the memorial in Nairobi. The organizations from which I had received a lot of help in the past were the obvious places to begin, but I was turned down by them all. They thought the idea of a building in Nairobi was excellent but they did not wish to become involved with such a large construction venture in Africa.

Fortunately the Phelps Stokes Fund, run by a former American Ambassador, Franklin Williams, was interested. This fund has done a tremendous amount of work in Afro-American relations and many African students have benefited from its support programmes. Franklin Williams, together with the Phelps Stokes board members, saw the development of the memorial in Nairobi as a novel approach to the extension of education within Africa and they accepted my arguments that there was great benefit in learning about the beginning of mankind. The origin of basic human behaviour, culture and language is a fundamental issue and a knowledge of the earliest phase of human evolution is important. With the backing of the Phelps Stokes Fund, I got busy with fund raising and as far as the U.S.A. was concerned, I began to rely increasingly on David Look to set up contacts. He was able to arrange meetings, luncheons and other events for me. David became the 'campaign manager' and in addition, he put friendly pressure on virtually all his own friends and neighbours as part of our drive for funds. I don't know how many friends David lost as a result of his association with me. I can only hope it wasn't too many.

I was very anxious to raise the financial support for the memorial from a number of different countries because I felt that I should not depend entirely on the generosity of Americans. By good fortune, I was able to achieve this. I successfully raised substantial sums from England, Sweden and Holland as well as from America. In the case of the Dutch, the contribution was equivalent to $200,000 and it was paid through the Dutch Development Aid Programme. With the Swedish grant, my initial contact was with the then Secretary General of the Royal Swedish Academy of Sciences, Professor Carl Gustaf Bernhard. The Academy has done a great deal to help in the development of scientific institutions in Third World countries and the prehistory of man was of growing interest in Sweden at that time.

As part of my effort to develop interest in the project in Sweden, I accepted an invitation to speak at the Royal Swedish Academy of Sciences on the topic of African prehistory and our origins. A special visit to Stockholm was arranged for this event but for some reason I became extremely nervous about the whole affair. Many of my colleagues who had learned of the invitation expressed amazement; in our discipline, to speak at the Royal Swedish Academy of Sciences was considered a very great honour. I had lectured hundreds of times elsewhere, often to very prestigious audiences but never before had I felt even slightly nervous.

Normally I speak entirely without notes but this time I decided to prepare my speech on paper. I was a little alarmed that, as the moment drew close, these meticulous preparations had no calming effect on me whatever. Dressed in a dinner jacket and feeling desperately uncomfortable and stiff, I was introduced to the various senior academicians and, as is typical of Sweden, there was a great deal of formality. My talk was to be held in the original meeting room of the Academy, an oak-panelled room whose walls were covered with portraits of famous scientists associated with the Royal Swedish Academy of Sciences. A large number of people were present and extra chairs had been added to accommodate everybody in what is a fairly small room. At the appropriate moment, the gathering moved into the meeting room and it was at this point that I was introduced to King Gustav who had just arrived to participate in the meeting.

After a few moments of Academy business, I was formally introduced to the gathering and then I made my way rather hesitatingly to the podium. I began with respectful reference to His Royal Highness and I acknowledged the great privilege that was being accorded me. With those brief remarks delivered successfully my mind went absolutely blank. I had no idea what to say next or how to extricate myself. I had my notes but they seemed to be irrelevant and I was simply quite unable to use them. I searched the room desperately hoping for inspiration and after what seemed an age, I found myself staring blankly at the King who seemed to be aware of my terrible dilemma. For some reason he smiled, perhaps more in amusement than in sympathy but that broke the spell. I put my notes away in my back trouser pocket, the only pocket in my dinner jacket outfit and gave a talk extemporaneously. It went very well and at the end I was acknowledged by a brief applause. This took me quite by surprise because I had been told that the academicians never applaud at a meeting. What had gone wrong? It turned out that the meeting had been changed from the normal closed session where only academicians were present to an open session where invited guests were included. The applause was therefore quite in order. After the lecture and a few questions, the gathering moved to the dining room where a sumptuous dinner was served along with the speeches and toasts that accompany a formal meal in Sweden. I was glad to have been thoroughly briefed on this aspect well in advance for it would have been only too easy to appear clumsy and insensitive to Swedish hospitality which, if understood, is very warm.

ABOVE My convoy breaks its journey in the boulder-strewn desert as we approach the southern end of Lake Turkana

BELOW An aerial view of our first camp at Allia Bay on the eastern shore of Lake Turkana

My father gave me invaluable lessons on the significance of various specimens that were collected from the Omo sites

CHAPTER TWELVE

FILMING, FINDING AND DATING

IT WAS NOT LONG AFTER MY LECTURE to the Royal Swedish Academy that I became involved with plans for a film on our work at Lake Turkana. There were several reasons behind this beyond simple ego. I was sure that a TV documentary would help me to raise public support for the memorial to my father both by promoting an interest in the science and improving my own credentials. In addition, I had asked a friend in England to help me raise money for the memorial and he had promised to get a substantial donation from his own company if I would help with TV documentaries in Kenya. This friend, Aubrey Buxton (now Lord Buxton), had been my guest at Koobi Fora in 1972 when Prince Philip had spent a week's holiday with me looking at birds and fossils. Prince Philip is well known for his love of Africa but, before 1972, he had not had the chance to see the east side of Lake Turkana even though he has a particular liking for that part of Kenya. Naturally, I was delighted to be able to arrange a private visit for him. Meave, who had given birth to Louise a few days before his visit, was unable to join us. Aubrey was very keen to have a film made by his company, Survival Anglia Ltd, for release on British and American television. In return for cooperating with this and other projects the Museum was to receive funds towards the construction of the building.

The making of a one-hour documentary for television calls for considerably more effort and energy than most people suspect and in this case matters were not made easier by the fact that I had my own definite ideas on how the film should be put together. Often I found myself at odds with the producer even though he was actually an old friend. Our friendship was often tested to its limits but I am pleased to say that it survived although the finished documentary was not as good as it should have been. The best way to describe the film is that it is good but obviously made by a committee. In England it was shown as the 'Bones of Contention', while the American title was 'The Three Million Year Old Clue'.

In June 1974, shortly before the making of the film, Meave gave birth to another daughter, Samira. The film attempted to document my initial visit to Koobi Fora in 1967 and from there followed the project through to the time of filming, highlighting major incidents and events. Neither I nor any of the rest of those involved were actors and it was terribly difficult to re-create excitement and enthusiasm for events long since past. The discovery of certain skulls seemed to be critical in our producer's idea of the story and never will I forget the time we spent 'finding' the complete *Australopithecus* skull in 1969 all over again. There must have been nine or ten retakes and by the end of it Meave and I no longer cared if it remained lost for ever.

There were other aspects of the filming that I recall very well. One particular idea was to show on screen how our early ancestors may have obtained meat by scavenging. It is my view that before hunting became a means of livelihood for our forebears, the stealing of meat from carnivore kills must have been a well-established practice. I am sure that bands of humans would frequently have been able to drive carnivores from their kill. The success of this would, of course, depend upon a good knowledge of carnivore behaviour which I am sure early humans possessed. To illustrate the point, our producer was eager to have me and a group of colleagues demonstrate how easily this could be done. Try as we would, we never found lions on a kill and eventually it was decided that we would cheat just a little. A scene of me chasing a lion would be intercut with a scene of a lion on its kill which, if necessary, could be obtained from library stock. The prospects of finding a lion to chase seemed better and the plan went ahead although I admit my enthusiasm was not tremendous.

One afternoon the lion-scouting party came back to say that a single lioness had been located asleep in a thicket of bush not too far from camp. This was the news that everyone was waiting for and soon all was ready and we set off. The first thing was to set up cameras on one side of the thicket so that I and my colleagues could approach from the opposite side. We were to flush the lioness from the bush and chase her past the camera, which was positioned so that whichever way she ran, we would all be in the frame. A 'white hunter' was positioned with the camera and he had instructions to shoot in the event of trouble. This idea upset me greatly partly because I had no confidence that the man could hit a target anyway (he was a great consumer of whisky) and also because I was to be approaching from exactly the direction in which he would have to fire.

There was too little time to argue and so I set off with my group of seven men to position ourselves for the scene. We had a walkie-talkie with which to maintain contact during the approach. As we neared the bush from our side, I came across very fresh tracks obviously made by the animal that we were approaching. A quick check showed that there were tracks of no less than five adult lions, three half-grown cubs and at least five tiny cubs! Quite a different proposition to a lone lioness. I made urgent contact with the producer and reported my discovery. My view was hotly disputed by the 'white hunter' who was adamant that we were dealing with only one lioness.

I decided then and there that the risks were too great to continue the approach so we all returned to the cars. The producer was furious, accusing me of cowardice to which I gladly acceded. I am sure that a family of lions with defenceless cubs would attack if provoked and I am also sure that our ancestors would have been much too wise to attempt anything so foolish. The filming came to an end and we never had another chance to shoot this scene – fortunately, I suspect.

Not long after this incident, I did have a chance to demonstrate that lions, if rushed at under the right circumstances, will leave a kill. One evening at Koobi Fora, after dinner, I suggested to some friends that for fun we should go down to the beach, which was about 100 yards from the camp, and try to catch a crocodile. At night, crocodiles' eyes shine bright red when illuminated by a torch beam and if you are quiet you can creep to within a few feet of them. Whether you then try to catch the crocodile is an individual choice, so far never taken at Koobi Fora! The fun, of course, is the stalk along the beach.

On this particular night, a large quantity of chianti had been consumed with the food and none of us was as alert as he or she should have been. Anyway, my suggestion of a crocodile stalk was received with enthusiasm and five of us set off for the beach armed with torches, some of which were brighter than others. From the beach, we cast our beams along the water's edge and there, indeed, were a number of pairs of bright-red glowing eyes. Our attention was focused on a particularly large pair of eyes that seemed to be well up on the beach away from the water's edge. Perhaps here was a chance to isolate a large croc from the water, and the prospect of some real sport was quickly grasped by all of us. I suggested that we make our initial approach with torches off and then from about thirty yards we should rush at the animal waving our lit torches and try to head it off from escaping back into the lake. My plan was agreed and we started our stalk in the dark. At a distance of about thirty yards, a quick use of my torch confirmed that our objective was still in place, very large and, upon reflection, rather un-croc-like. There was no time to consider this because the last rush had begun. At about ten yards our rather dull lights showed a large lion hurriedly departing from a zebra which was lying on the sand.

The lion had been in the process of killing the zebra when we made our final attack! The poor zebra had a broken neck but was not yet dead so we had to kill it to release it from its misery. The lion was so put out that it never returned. We gradually realized how silly we had been; back in camp, over coffee, we agreed that crocodile-spotting parties should carry bright torches in the future. I am appalled to think of what might have happened had the lion chosen to stay on its kill.

Kamoya's team regularly used to scavenge from lion kills when there was a shortage of meat in camp, and often this meant chasing lions from the kill, but this was always done with a Land-Rover. If a lion was heard roaring during the night, the next morning would start with a search to locate it. As often as not the lion would be at a kill, either alone or with other lions. Depending on the type of animal killed and the amount of meat remaining, Kamoya would drive the vehicle towards the carcass, so driving off the lion or lions. Several team members would then get to work to cut off a fair helping of meat before allowing the lion to return. The principal problem was that in most instances, the car used was quite open, without a roof, and offered little in the way of protection.

On one occasion, Kamoya's team had decided to take the whole kill because not much had been left by the lions. Having got the carcass on board, they drove to where they were working for the morning, planning to return to camp with the meat at noon. On the way back they met the lion which, quite reasonably, was angry and gave chase. The car was grossly overloaded and because its engine was badly tuned the lion very nearly got close enough to leap into the back where there were eight terrified men and one rather fly-covered carcass. By good luck, aided by considerable encouragement from the team, Kamoya kept ahead of the lion which eventually gave up the chase. Since that day, kills are only taken after the lions have finished and left the scene!

Nevertheless, I am quite convinced that humans can scavenge successfully. The fossil record from sites in different parts of the world suggests that it was probably only during the past million years that the deliberate hunting of large animals became a characteristic of human behaviour. Prior to this, our ancestors may have hunted rodents and reptiles but if they ate any quantity of meat it must have been obtained by scavenging. We know from archaeological remains that from about two million years before the present, early humans were using cutting tools made from stone and eating increasing quantities of meat. Sometimes the bones of animals that have been devoured by predators show tooth marks and it is now possible to distinguish these from the cut marks made by stone tools. If the cut marks are superimposed upon the tooth marks, then there is a good chance that the hominids were scavenging from the kills of other predators. It would be very interesting to know when it was that humans began to be feared by the animals that they hunted. Presumably, in the scavenging days, man and the other animals mixed freely, just as baboons today mix freely with antelope, zebra, wildebeest and a host of other plains game. Perhaps it was gunpowder and the gun which fundamentally changed the relationship between man and wild animals. It is interesting to observe that when a predator, such as a lion, is hunting, prey animals that are aware of it will simply move aside, keeping it in sight, with no panic. Once the hunter has been seen, it cannot take its prey by surprise and so it is not, generally, a threat – but a man with a gun is.

Before we began filming, and at the time the field season was just beginning, an incident occurred that I sincerely wish could have been avoided; it emphasized to each of us how easy it is to get into trouble in the field. One of the projects for the 1974 season was the collection of soil samples by a professor and a graduate student. The study of the soil chemistry was part of an attempt to reconstruct the palaeoecology of an area that had been occupied by our ancestors more than one and half million years ago. The samples had to be collected over a wide region and to do this the geologists were operating from small camps. There is a long-established rule on my projects that no person should work alone in the field. I believe that if workers operate in pairs there is a far better chance that accidents will not happen but if they do, one of the pair will be able to offer assistance or call for help. A lone person can quickly get into trouble which, in this part of the world, can often be fatal.

The incident that I relate was tragic. The graduate student, an American, whom I do not propose to name, went off to collect samples and for reasons that I still do not understand his colleagues permitted him to go alone. He was to go to a site where he and his colleagues had worked together the day before and the distance that he had to walk was less than a mile. He was given an aerial photograph on which the site was clearly visible and as he left, shortly after breakfast, he promised to return at about eleven o'clock that same morning. The other scientists at the camp made a mental note of this and went off during the morning to do their own things. At midday, when people returned to the small camp for lunch, it was noticed that the student was not among them. They were slightly concerned, but I suppose thought that he had got something rather interesting and would be back shortly. They therefore decided to have their lunch and not worry.

By two o'clock in the afternoon when the student had still not returned, those in camp realized that something must have gone wrong. Nobody would willingly remain out in the heat of the day. A search party was set up, its members going first to the site where the student had gone to collect samples. There they found recent footprints and evidence of digging. Where was he? There was not a sign. They called, and they whistled, and they walked in various directions but after an hour or so they grasped that whatever had happened he was not easily to be found. The decision was taken to call in help, and a car was despatched to the northern end of the area where Kamoya was working. He was reached at 5.30 in the evening and he immediately set out accompanied by his men to join in the search. By that time it was almost dark, so they continued the search using car headlights and by lighting fires on ridgetops until about nine o'clock but with no success. Everyone was distraught. Kamoya rushed back to the base camp at Koobi Fora and sent a message for me to come and join the search with the aircraft. I left Nairobi at the crack of dawn the next day to find that Kamoya had also brought in a missionary plane that was nearby and by midday a full-scale search was under way involving seventy people, some of whom were members of the National Park Ranger Force. We combed the area back and forth, making an increasingly wide circle of activity. Although, initially, we had picked up footprints working towards the lake which was to the west, these were soon lost in the stony ground. We had no idea what had happened.

By the end of the fourth day the search party was exhausted. It was increasingly unlikely that we would find the student alive because in the tremendous heat of the desert it is not possible to survive long without water. It was more and more probable that the exhausted searchers would get themselves involved in a silly incident that would result in a further tragedy. I decided that if by noon on the fifth day we still had not found the student, I would have to call off the search. All sorts of explanations for his disappearance kept running through my mind and I thought that the most likely was that he had run into trouble with a large carnivore, perhaps a lion, and he had simply been killed and eaten.

The search was about to be given up on the fifth day, on the morning of which I flew back to Nairobi to confer with the American Ambassador. After I had left for Nairobi and while the members of the search party were dispersing, Abdi Mohammed, the warden of the National Park, was driving back along a perimeter road when he noticed a pile of stones in the middle of the track. Beneath the stones was a diary and after a quick search the young man was found under a nearby tree. He was barely alive and delirious so it was impossible to find out exactly what had happened. He was found so far from the area where we had been looking that it was obvious that he had got lost very early on and had walked completely out of the zone where we had expected to find him.

Although alive he was very ill and having radioed Nairobi for medical advice we administered first aid. I then flew him back to Nairobi where he was admitted to the intensive care unit of the city's hospital. Tragically, the effect of dehydration over the five days was too great and the doctors were unable to save his life. It was a terrible shock to us all and emphasized the vital importance of taking constant care when operating in the desert – it is only too easy to be lulled into a sense of false security when things are going well. The loss of this student, the expense of the massive search, and the near tragedies that we had as a result of the extended quest, had a very sobering effect on my management of the expedition. Since then I have become very critical of the way things are done in the field and I insist that anyone working on any project of mine follows the rules without discussion.

In February 1975, I attended a symposium in London that was organized by a good friend, the late Bill Bishop. The theme of the meeting was to discuss recent scientific work in the East African Rift Valley and it included contributions from a wide range of colleagues and friends who were involved with work at Olduvai, Baringo, the western and the eastern shores of Lake Turkana and the Omo Valley. It was an excellent meeting and the results were published some years later.

The meeting was held over two days in the rather imposing premises of the Geological Society, which are in part of Burlington House, in Piccadilly. I remember the meeting well because it marked the first major public row on an issue that was to become a running sore over many years. I hope that it is now far enough in the past for me to recall what happened at the meeting objectively although I realize that in presenting my version I am probably a little biased. The subject of the heated exchange was the real age of the KBS tuff, the horizon of volcanic ash in which we had found our alleged 'earliest tools' in 1969, and the horizon which overlies the deposits from which we had recovered the famous '1470' skull that I had so proudly announced at Lord Zuckerman's meeting a few years before.

One of the speakers at the meeting was Frank Fitch who had dated the KBS tuff and who was presenting a joint paper with his colleague John Miller of Cambridge. Their contribution was to present new and improved data confirming the age of the KBS tuff to be 2.4 million years old.

They allowed a 10 per cent margin for experimental error but this was of little relevance. Their date was further supported by one of their students who had dated the tuff by the fission-track method.

Before Frank Fitch spoke, the meeting heard a number of papers on the excellent work of Clark Howell's group in the Omo. One of these was by a scientist, Basil Cooke, who had spent much of his working career in Canada. At the time, he was unquestionably the world's foremost authority on the fossil pigs of Africa. I had come to know him as a colleague of my father, and he had been a frequent visitor in Kenya throughout my childhood and adolescence. It was therefore perfectly natural for me to have invited Basil to 'do' the Koobi Fora pigs for me. I had hoped that Basil would describe the excellent fossil material from the sites and I expected to see a draft paper which would subsequently appear as one of the publications of the group working at Koobi Fora. However, I doubt if I made any of this clear to Basil.

My invitation to Basil had been made in 1969 and he had made several visits to the National Museum in Nairobi to measure, photograph and study some of the quite exquisite fossil pig skulls that we had collected since 1968. Some of these fossils were the first complete examples of species that had previously been known to science only from fragmentary material. Until the time of the London meeting, Basil had not shown me his draft which, as leader of the Koobi Fora Project I felt was my due. I had, in fact, begun to worry whether or not he was giving adequate attention to our marvellous finds. I needn't have worried because when Basil got up to speak at the meeting, he gave an excellent paper which made reference to fossil material from several sites including Koobi Fora. The problem that concerned me, however, was that he used the pig data, including our own, to question seriously the validity of Frank Fitch and John Miller's dating of the KBS tuff. He suggested that on the basis of the comparisons he had made of fossil pigs from other dated sites, the KBS was in reality about 1.8 million years old. At the time I was both taken aback and upset because I felt that Basil, as a member of 'my' team, should not have used the Koobi Fora data in the way he did without giving me a full report before the meeting.

In fact, in 1975 I was well aware that there were a number of serious questions about the validity of the 2.4-million-year date. In 1972 Vince Maglio had raised doubts about the dating based on his study of the fauna in general and the fossil elephants in particular. And in 1973, Basil had spoken at a conference in Nairobi where he certainly expressed his reservations. I had also learned from colleagues that Frank Brown and Garniss

ABOVE It took time to learn how to persuade our camels to do as bid!

LEFT The first permanent structure at Koobi Fora which we used as a store between expeditions and, in 1970, as a home

In 1975 Bernard Ngeneo noticed the exposed brow ridges of a human skull *(left)*. After an initial examination, I carefully excavated it and removed the fossil intact. Kamoya Kimeu watched *(below right)* while I completed the final excavation

ABOVE Our field camps are simple and we always cook on open fires

LEFT Sunsets at Koobi Fora are often very beautiful

Curtis, experts in radiometric dating who had worked on the geology and dating of the Omo, were seriously concerned about the presentation of data that Fitch and Miller had put forward in support of their KBS dates.

At the time of the London meeting I was, nevertheless, still impressed by the work of John Miller and Frank Fitch. As a consequence, I set out at the meeting to defend their date for the KBS tuff; I was probably less than subtle in my remarks and I am quite sure that I offended a number of people. In any event, I did point out that while pigs and pig evolution were important, I was far from certain that they could be used in the way that Basil proposed. I also noted that nobody really knew enough to judge whether Basil was right or wrong because at that time only he possessed detailed knowledge of pig evolution in Africa. Glynn Isaac was also at the London meeting and by then, too, he had some reservations about the Fitch and Miller date, following some independent work by a number of American colleagues. Nevertheless, we both felt that there was a long way to go before we could abandon the 2.4-million-year date ascribed to the KBS tuff. As the joint leaders of the Koobi Fora Project we sought to reserve our position pending further research, which we planned to initiate.

One of the consequences of this meeting was that I encouraged John Harris, who was then Head of Palaeontology at the National Museum, to undertake an independent study of the fossil pigs from Koobi Fora. I wanted our material to be fully studied, described and published, and I also hoped for an indication as to whether it was possible to use pig teeth in the way that Basil had proposed. John agreed to do this and shortly afterwards, I invited Tim White, a young graduate student from Michigan University, U.S.A., to work with John on the project. At the time he was studying anthropology and was anxious to have access to some of our fossil hominid specimens for use in his thesis.

Tim and John subsequently devoted almost two years to collecting and compiling their data and eventually produced two papers which effectively confirmed everything that Basil had proposed at the memorable London meeting. It was unfortunate that for a variety of reasons Basil had not been able to be directly involved with the Harris-White study because it was always our hope to produce a three-authored paper. Often, in science, issues of the kind that I have described lead people towards new ideas and tighter controls. Sadly, schisms and broken friendships also result. I certainly regret that, as a result of my determined effort to see Koobi Fora and our work in Kenya given its due, I lost several friends. Naturally enough, as the saga of the KBS tuff continued, there were further meetings and arguments and it was only in 1980 that a broad consensus was finally achieved. For my part, at the time of the London meeting, I was aware of the possibility that Fitch and Miller might be wrong, but not knowing anything about geophysics and dating, I was in no position to make a sensible judgement and instead I was vigorous in their defence. Nevertheless, as John and Tim progressed in their studies, I was kept informed and I became aware of the increasing evidence against the 2.4-million-year date.

As a result, Glynn and I decided we should invite other geophysicists to work on the KBS date. Eventually we managed to arrange for several different laboratories to evaluate the same material from split samples, using two methods; fission-track dating, as well as conventional potassium-argon. This was done quietly and with little fanfare. As a result it became quite clear that the KBS tuff is no more than 1.9 million years old. The fossils found below that horizon are, of course, older but just how much is impossible to say and it would be prudent to think of the skull KNM-ER 1470 as being about two million years old. This reduction in the age of '1470' did not, of course, alter the fact that this large-brained hominid had been found in deposits that had also produced *Australopithecus*. It still demonstrated beyond doubt that more than one type of hominid had existed at the same time.

A positive aspect of all this is that the KBS tuff must now rank as one of the most securely dated deposits in Africa and the drawn-out debate has certainly resulted in a general improvement in the science of dating.

In 1975, Bernard Ngeneo made what is perhaps, so far, the most significant find of all at Koobi Fora and at the time Meave wrote a fairly detailed account of the circumstances. It is interesting to quote directly from her diary:

August 1st. Friday. Richard decided to stay in camp all day as he is very tired after a busy week in Nairobi. We were also expecting some Japanese visitors but they did not appear. When Kamoya and his team returned to camp they reported that Ngeneo had found a hominid brow ridge and maxillary fragments in Area 104, probably just above the KBS tuff. Our curiosity was roused when we asked Kamoya if the brow ridge was attached to a skull and he replied 'Well we cannot say because we can only see a very small piece, most of it is buried.' Might it be a complete skull?
August 2nd. Saturday. I went with Richard in the Toyota to see Ngeneo's brow ridge following Kamoya and the team in their Land-Rover. They

showed us the site on the slope of a small gully. There were two fragments of maxilla on the surface and the frontals and the top of the orbits were just beginning to erode out. It is impossible to say how much is there. The others all went off not wanting to do any damage. Richard began to excavate the skull slowly while I took photographs and collected the maxillary fragments and numerous other tiny pieces of bone on the surface.

The skull is in very bad condition with small plant roots growing through numerous tiny cracks, but it does seem to go on back. Richard exposed part of the parietals and then left it to dry and harden. We returned to camp and arrived just in time to welcome the Japanese party and give them lunch. After lunch Richard went back to the skull with Kamoya and exposed some of the temporals and one ear hole. It is an extremely slow job as the skull is so fragile and the ground is wet from recent rain. There would be nothing left if it had eroded out naturally; Ngeneo found it just in time, though how he spotted it I cannot imagine. There is some matrix on it but the main problem is the numerous tiny roots which get into the bone and break it up . . .

In the evening we covered the skull with an upside down metal basin and some thorn bush before returning to camp. It would be terrible if an oryx trod on it now!

August 3rd. Sunday. Day in camp. Light rain. Rather cold.

August 4th. Monday. I went with Richard to the hominid skull. He worked on it until 10.30. It looks as if it has a large cranial capacity and it may well be complete. It is a very slow and difficult job though. We then went to see where Kamoya had set up his camp. He had left with the team early in the morning with enormous loads on the cars, but they had arrived safely. Richard went back again after lunch to continue his excavation. Weather at last seems to be more 'normal'.

August 5th. Tuesday. We returned to the hominid site and as Richard worked the skull began to look all that we had hoped and very exciting . . . It is in very bad condition with roots growing into the bone and everywhere the bone is broken into tiny pieces which by some miracle still remain in their original positions. Richard really has remarkable patience for this type of excavation, and this specimen certainly needs it.

Richard went back to the skull after lunch but he could not do too much because it is so hot. We saw Kamoya and arranged for some of his team to start sieving the hominid site in the mornings . . .

Richard seems to have a temperature on and off, he puts it down to too much sun while excavating. He has to fly to Nairobi tomorrow and has decided to persuade Bob Campbell to come back with him to film the final stages of the excavation.

August 9th. Saturday. Richard arrived with Bob Campbell at about eleven o'clock. After lunch we went to take out and film the skull. It came out well with no disasters and the bone is now hard. It looks amazingly complete. An added bonus is that the sieving has produced some teeth . . .

What a find it was! There was no doubt that this was not *Australopithecus* nor even *Homo habilis* but rather *H. erectus,* a more immediate ancestor of ourselves. Words are inadequate to describe our feelings because for months we had suspected that *H. erectus* had lived in Africa more than a million years ago and here at my fingertips was proof, a perfectly preserved skull found *in situ* in sediments over 1.5 million years old. This specimen is one of the oldest-known examples in the world of a skull of *H. erectus,* but earlier traces may yet be discovered.

Meave's account illustrates well why I have made it a rule that all fossil hominid finds at Koobi Fora are left undisturbed until I can personally visit the site, assess excavation strategy, photograph and record the specimen and make the collection myself. The fossils are often extremely fragile and it is essential to treat them with great care. Often the fossil is badly broken up with a number of small fragments lying on the surface of the ground. These, and other pieces that may be partly buried under the top soil, have to be collected and glued together with adhesive and hardener on the spot. All this takes time. In the event that a specimen is partially buried and it is not certain whether or not it is hominid, the earth covering has to be removed with extreme care. There have only been three finds in nearly 200 where this rule has not been observed and as a result the specimens damaged. In each case, the offender was an outsider from abroad working with Kamoya's team. Presumably the individuals felt that their academic credentials absolved them from working strictly within our regulations.

The most unnecessary incident of all was when an American graduate student decided to check a particular find that one of Kamoya's team had reported as a possible hominid. Kamoya directed that it must be left undisturbed until I had arrived. The student was impatient, ignored Kamoya's ruling and, with the aid of a large hunting knife, he levered the bone from the ground, breaking it in two places. It was immediately apparent that the bone was not simply a rib of a large animal as he had suggested but a well-preserved shaft of a hominid femur. He quickly reburied it but the damage was done.

Once safely back in camp at Koobi Fora, the new *Homo erectus* was carefully packed in a suitable box for the return journey to Nairobi in our small aeroplane. Before leaving, I took a series of photographs of the skull and left the film in camp as a record should the flight to Nairobi end in disaster. Normally I do not give a thought to such matters but on this occasion I was carrying the most perfectly preserved skull of *H. erectus* ever found, let alone the earliest, and I felt only too aware of my responsibilities. In fact, our flight was uneventful.

One of the problems with the new find, which now bears the catalogue number KNM-ER 3733, was that the inside of the brain case was completely filled with very hard calcified rock. This made the fossil extremely heavy and difficult to handle and had it fallen, the impact would have shattered it. In addition, it was quite impossible to think of reconstructing and attaching the delicate facial bones to the skull because if the specimen were handled carelessly the excessive weight would certainly crush them. How to remove this solid rock from the braincase posed quite a problem. Normally, fossil skulls are found broken and so the question does not arise or alternatively, if the specimen is relatively robust it is considered reasonable to leave the rock inside the braincase. One such example is the *Australopithecus boisei* skull that Meave and I found in 1969. This is still filled with rock, although some day it will have to be dealt with.

Fortunately, in the case of the *Homo erectus* skull, I had the assistance of Alan Walker who, based in America, came out to work on a number of our fossils in January 1976. Alan has extraordinary skills and has devoted months of his life to the tedious business of cleaning the rock off our more precious and fragile specimens. The challenge presented by KNM-ER 3733 seemed to appeal to him and he was soon spending a great deal of his valuable research time working on the specimen. I was very curious to know how he proposed to remove the rock from the inside of the skull, but I decided to keep my silence until he was ready to tell me. His plan of action, when he eventually disclosed it, came as quite a shock. He wanted to split the skull into two or more large fragments by driving a large cold chisel into the rock-filled braincase with a heavy mason's hammer!

Alan went on to explain his plan in detail. First, he was going to cover the outer surface of the skull with a number of layers of tissue paper, each bonded on with a water-soluble glue. This was to produce a firm casting that would prevent the skull from bursting or shattering when the chisel was being driven in through the small opening at its base, the foramen magnum. The rock in the immediate vicinity of this opening was to be drilled away as far as was possible, making it easier to use the chisel. It all sounded perfectly sensible but I confess I had terrible qualms. What was to happen if the skull did shatter? 'Well,' said Alan, 'at least all the pieces will be inside the paper lining and we will simply have to put them all together again.'

Over the next day or two the skull was duly covered with its paper jacket and left to dry. Several days later Alan was ready and, assisted by Meave and Tim White who was then working on our fossil material for his thesis, he set to work. I preferred to be away from the Museum for the whole day. The skull was firmly set down on a sand-filled bag while Alan operated the mason's hammer and chisel. Blow after blow and nothing happened. Alan was exasperated and finally put the chisel to the stone wall on the verandah and with a very moderate blow he knocked off a huge flake of stone! A larger chisel was purchased from a hardware store in town and soon after Alan had obtained the result that he had expected: the skull broke into three pieces and the breaks were absolutely clean. Several weeks later, after Alan had removed the rock from the inner surfaces, the pieces fitted back together again perfectly and today there are no external signs that the fossil was subjected to such extraordinary treatment. Even though I was not present I well recall the agony of wondering what was happening to the specimen, and it is perhaps because of this that I cannot yet bring myself to allow the *Australopithecus boisei* skull to be so treated.

This *Homo erectus* skull has become one of the most significant discoveries from Koobi Fora because it represents the earliest record of this species in Africa. Although there is some claim for even earlier material in Indonesia, I am personally unconvinced and believe that '3733' along with other African evidence for the ancestor of *H. erectus* make it clear that *Homo* was an African development. I believe that it was *H. erectus* that developed the technological capability to get beyond Africa some 1.5 million years ago.

Chapter Thirteen

FIGHTING FOR TIME

During the latter part of 1975, I spent most of my time in Nairobi where my primary concern was the Louis Leakey Memorial Building. At the time we certainly did not have enough money to finance the project, no more than $150,000 in fact, but I was reasonably optimistic that additional sums would be forthcoming from various foundations and friends as soon as the project was under way. I was worried by the ever-rising estimates for the cost of construction. The effect of oil shortages had not yet been appreciated but prices were rising fast in Kenya. Because of this and because I am a strong advocate for working under pressure, I decided to go ahead and have a contract prepared for the construction of the building. In 1976 I signed a contract for a total of just over $US1,000,000.

Once I had signed, it was not long before the builders were on site and work was in full swing. This provided a tremendous incentive for me to raise the additional money and I began to spend more and more time away from Nairobi visiting friends abroad and making appointments to see foundation and government people all over the world. It was an extremely interesting time and I learned a great deal about people in high positions and the functions of bureaucracy. Although I found that a very wide range of people have a basic and fundamental interest in the subject of human origins, I was surprised to learn how cynical and doubtful people were about the stability of Africa as a whole. One of the things that most amazed me was that so many intelligent and well-educated people look at Africa as if it were a single country. Africa is at least three times the size of the United States of America and it has forty independent countries with an immense variety of climate, terrain, peoples and so on. The fact that there may be instability in one area at one time should not be interpreted as applying to the whole continent, yet I personally experienced a great deal of difficulty explaining that Kenya was a stable, peaceful and progressive nation. I had frequently to explain that events in South Africa and Uganda, for example, had absolutely nothing to do with what was going on in my own country.

While raising money and talking to people about my plans for the memorial to my father, I found that many were suspicious of my motives.

Many of the scientists I talked to about the goals of the memorial were willing to give their support in principle, but they had reservations about supporting me personally, perhaps because they thought I was trying to build a private power base.

In fact my objective was simple enough. I wanted the National Museum to have a building in Kenya where the large collections that had been made by my father and others could be safely stored. I wanted to build new laboratories, a good library, facilities for technical work, a dark room, X-ray facilities and a maximum security area for the storage of some of the more critical fossils of our ancestors. I also wanted to attract international funds in support of research programmes. The Museum's Board of Trustees decided that instead of running the memorial simply as a department of the Museum, it should be established on a semi-autonomous basis. We decided to create a Memorial Institute that would operate as a subsidiary of the National Museum. The concept was that the National Museum would own the Institute but that the latter would be given a degree of self-management which would enable it to attract and participate in international activities. It was also helpful for me to be able to talk about an institute that was slightly separate from the Museum because I could then demonstrate that although I was raising the money to build it I had no intention of being directly involved with its day-to-day management.

Before long we were actively engaged in a search for a director to run the Institute which was, of course, still in the process of being built. I was anxious that whoever we employed would be a mature person who could look at both the national and international needs of the study of human origins. We were anxious to explore the possibility of establishing international training programmes and making the Institute a centre for east and central African studies. There was no pressure on us to employ a Kenyan in this position. The director could be from any part of the world though there were advantages to having somebody familiar with the ways of the country in which the Institute was to operate. Accordingly, it was decided that we offer the job to a senior Professor of History at the University of Nairobi. Professor Alan Ogot, whom I had known for a good many years, accepted our offer of the post on 1 July 1977. I was very pleased to have another person responsible for developing plans for the Institute, because as the completion of the building approached I was increasingly unwell and found that I had too little time and energy to attend to all the minutiae. The last decision I made concerning the construction of the building was to tell the contractors to go ahead and

ABOVE Meave and I always enjoyed collecting the fossil hominids found by Kamoya and his team

LEFT A preliminary cleaning of the *Australopithecus* skull in the field, near where I found it while exploring with Meave close to the Kenyan-Ethiopian border

ABOVE Our favourite home at Koobi Fora was totally destroyed by fire

LEFT The subject matter of the after-dinner discussions at Koobi Fora, which are often lively, vary according to which scientists are in camp build a 500-seat auditorium as an adjunct to the memorial. There were then no auditoriums of any size available in Nairobi and I believed that such a facility would be of tremendous benefit to the National Museum. Having the contractors already on site for the construction of the Institute made it much cheaper to go ahead with this further building. I did not have any funds for the auditorium but I was reasonably confident that I would be able to raise the money within a year or two of its completion. In the meantime the money would have to be borrowed. What I had not fully realized was that my health was deteriorating so fast that I was unable to spend the necessary time to raise the funds.

In early 1977 the Museum Board decided that the memorial would be formally called the International Louis Leakey Memorial Institute for African Prehistory. Construction was completed in time to allow for an opening ceremony immediately prior to a large international meeting, which was to be held at the Museum, on African Prehistory – the 8th Pan African Congress. This was especially suitable for many reasons, particularly because the very first congress on prehistory had been organized by my father in Nairobi in 1947. The inauguration was a great occasion with people coming from all over the world. There was great goodwill from many quarters and, to my pleasure, the success of this new venture seemed assured.

My work at Lake Turkana was slowing down during this time. I was becoming increasingly despondent about the amount of time I spent organizing the expedition. I seemed constantly to be worrying about money, the maintenance of vehicles, the deterioration of the camp and the need to repair buildings. I had less and less time to do what I enjoyed – to go into the field and look for fossils. Partly as a result of this, but more particularly because of the wealth of data that had not yet been fully written up or studied, I declared 1976 to be a slack year at Koobi Fora. I insisted that any scientist wishing to work in the field would have to be self-sufficient. I was not prepared to go on being simply general manager of this vast project.

During 1976, I became involved for the first time with publishing. I had for long been interested in writing a book and partly as a result of my lectures overseas and my attempts to raise money for father's memorial, I had met a great many people who encouraged me to put down some of my ideas about the subject of human origins in a popular way. I was tremendously busy with running the Museum, raising funds, and organizing the expeditions, so I knew I would not have the time to sit down and attempt any extensive writing. At the time, I was also very concerned about my health; I knew that I was sick, my poor kidney-function was a continuing problem which made me constantly aware that in the not too distant future I would have to undergo medical treatment involving considerable expense. I had no idea whether the treatment I required would be possible in Nairobi or whether I would have to travel abroad. If the latter, would I be able to afford the treatment?

Because of these problems the idea of writing a popular book about human origins was quite attractive. Whenever I received an invitation from a publisher to write a book, I always responded with interest if not commitment. It was in reply to one such letter that I received an invitation to meet George Rainbird of the Rainbird Publishing Group in London. His suggestion was straightforward: as I did not have time to write a whole book myself, I could get somebody to write it with me and we could share the proceeds on some mutually acceptable basis. This was sufficiently interesting for me to accept the idea of further discussions.

The outcome was a decision to publish a book which was eventually entitled *Origins*. As collaborator I chose Roger Lewin, a zoologist I had met when he was a science reporter in London at the time of the '1470' press conference. The collaboration proved to be the beginning of an excellent friendship. Roger spent a lot of time discussing ideas and he did a great deal of the essential research along with the initial writing. My involvement was to provide further ideas and suggestions on the book's structure. *Origins* was first published in 1977 and proved extremely popular. It was translated into ten languages and sold well over 500,000 copies. Following *Origins* I was subsequently involved in *People of the Lake*, a second book (with another publisher) that I had less to do with because of my deteriorating health. Again, Roger was my co-author. In retrospect I realize that had I not been unwell and worried about my family's welfare I would not have attempted to write a book as soon as I did. The most extraordinary aspect of working with Roger was that we never argued and have remained on excellent terms.

Once the Memorial Institute was opened with Alan Ogot as the full-time Director, I felt greatly relieved. I no longer had the responsibility of dealing with the day-to-day administration of a large part of the Museum that had previously been under my control. Many of my staff had been transferred to the Institute and the responsibility for safeguarding the archaeological and fossil collections and dealing with the various visitors was firmly handed over to Alan. Whether it was because I had less pressure, or because my health really was failing, I do not know, but certainly in the early part of 1978, I became aware that my condition had worsened. I frequently suffered from very severe headaches, some of which were so extreme as to make it impossible for me to do a normal day's work. The headaches were a consequence of my extremely high blood pressure and although I was under medication, it was not enough to keep me even reasonably well.

One of the things that I particularly wanted to do before I became too unwell was to organize a new expedition to search for some of the very earliest stages of human evolution. I was extremely interested in following up some work that my father had initiated with my younger brother Philip in an area south of Lake Turkana called Nakali. This is a very rough and rugged part of the Rift Valley where aerial surveys had shown a series of promising outcrops of sediment. The geology of that part of the Rift Valley suggested that some of these sediments would be perhaps as old as seven or eight million years and could, therefore, have rather crucial fossils relating to the evolutionary stages leading up to the appearance of the first hominids or bipedal apes. My interest was to establish the time of origin of our early ancestors; in other words, to find when the split between a four-legged ape and a two-legged human ancestor had occurred. It had long been thought that this must have happened sometime between four and ten million years ago but there had always been a shortage of evidence from that period. In any event, I felt that I still had the physical strength to undertake such a project and without the responsibility of having to run the prehistory section of the Museum and with less pressure to raise funds, I was able to plan to be away for sufficient time to investigate thoroughly this new locality.

I raised some funds for the project and in July of 1978, accompanied by Kamoya and his team, I went to Nakali. I had invited my good friend Alan Walker to accompany me and between the two of us we hoped to investigate in greater depth some of the localities that Philip and his party had looked at some years before and also to survey some sites he had not had time to reach. Our expedition had to travel with an armed escort, because in this part of the country there were bandits – known in Kenya as Ngorokos. The area is without roads and practically inaccessible so that any banditry is very hard to control. The Ngorokos, who are primarily concerned with rustling cattle, always herd into this part of Kenya, where they are reasonably safe from the long arm of the law.

Our expedition to Nakali was a great deal of fun and I was able to relax much more than I had previously been able to do. Despite this I was not able to do nearly as much as I had hoped; physically I was just not up to it and I found long walks and climbs particularly difficult. In addition to the high blood pressure and constant headache, my legs tended to swell very easily from fluid retention, all of which made it very difficult for me to keep up with the rest of the team. I made every excuse I could to fly back to Nairobi to collect extra supplies or to attend unimportant meetings leaving the day-to-day exploration and the arduous task of climbing the steep hills to Alan, Kamoya, and their team.

During the Nakali expedition my career nearly came to an abrupt end. We had set off for a certain area by car and on the way we had run into an extremely heavy rainstorm. As a result of the downpour, the open plains that we had planned to drive across in the dust were turned to mud and flowing streams. I experienced considerable difficulty driving the Land-Rover. The only way to get to our destination was to drive as fast as possible so that the extra speed carried the vehicle across the many slippery and boggy places. It was a hair-raising drive which, although quite safe, caused a great deal of excitement. Kamoya and the team were shouting encouragement all the way and by the time we arrived some two miles from where we had started, everyone was in high spirits chatting excitedly about their interpretation of certain corners and tight turns that I had had to make to keep moving.

The rain had stopped and we set off along the footpaths to spend the day exploring and searching for fossils. We were walking along in single file and there were several people behind me including Kamoya and Alan. Everyone was talking in Kiswahili and I was dimly aware that somebody was saying 'Go, Richard – Go – Go, Richard.' I thought that this was simply another anecdote about how someone had been encouraging me to drive as fast as I could. I heard 'Go, Richard' yet again and it seemed to be slightly more urgent and immediate. I turned to look behind me and there was a large cobra chasing along the path with its head reared up about to strike at the back of my leg. It turned out that Nzube, who was behind me, had seen the cobra come out on to the footpath just in front of him. Because he was blocking the path, the snake went along the footpath after me. Apparently it had been striking and missing the back of my leg by just a few inches while I, unwittingly, had misunderstood my colleague's warnings. Had the snake bitten me, I am quite certain that in my poor health I would not have survived.

The site at Nakali turned out to be important, but the few occurrences of fossil bones are high in the hills, often accessible only after a very difficult climb and although the finds were interesting there were no discoveries that warranted a second major expedition by me. Some rodent fossils that we found suggested that there was a close faunal relationship between this part of Africa and Asia about nine million years ago. I am quite sure that in years to come others will work Nakali to great scientific profit, but I think that it is unlikely that human or human ancestor remains will be found there without the expenditure of much time and money. Because of deteriorating health a second expedition to Nakali was certainly out of the question for me.

During 1978 I had discussions with a number of colleagues about the prospects of making a major television programme. In particular, I had talked to Graham Massey and Peter Spry-Leverton, both of whom were then working for the BBC as producers. I had first met Graham when he had visited Kenya to film a short programme called *The Ape that Stood up*. We got on well together and so, through him, I proposed to the BBC that we should do a TV series on man's evolution. I was very anxious to participate in the production of a serious film pulling together the whole mass of information on human origins that was now available. I was familiar with all the films that had previously been made in the Western world on the subject but I felt very dissatisfied with their treatment of what is a fascinating story. As a result of my initiative and the interest that Graham and Peter showed, the BBC was persuaded to look for the finance to produce a series. By late in 1978 we had an agreement from the BBC to make the film: a preliminary budget had been set and work had begun on establishing an outline for six one-hour programmes. Late in 1978, I invited Graham Massey, Peter Spry-Leverton and Roger Lewin to Kenya for detailed discussions about the content of the series and we spent a very useful week at Lamu on the coast.

It was clear that there was tremendous potential for us to make a superb film. It was decided that I should be the presenter appearing in each programme, while much of the detailed scientific information would be put across by selected scientists whom we hoped to film in their places of work. Graham had some reservations about my ability to perform before a camera and I must confess that I was a little uncertain myself. What I did not do was tell Graham that I was ill; I hoped that he was unaware of this fact. Indeed, I was determined that my illness should be kept a secret because it seemed certain that if the people at the BBC knew how sick I was they would never have expected me to undertake the arduous travel that was necessary to complete the project. Graham, apparently, knew that I was not in perfect health although he has since told me that he was not sure what the problem was.

In early 1979 I had to travel to America to give some lectures and participate in some fund-raising events. When I returned through London at the end of March I saw the BBC and we planned a reconnaissance trip to Europe before beginning the filming. The dates were such that I had to go back to Kenya before meeting them in France in June to begin the serious work on the preparation of the television series, which subsequently became known as *The Making of Mankind*.

When I arrived home I began to feel extremely unwell: fluid retention, headaches, high blood pressure and a general feeling of malaise resulting from uraemia, had all taken its toll. I found the days extraordinarily long. Whenever I sat at a table, if I did not get my feet up level with my seat, my ankles would swell. There were many other symptoms that made me quite sure that I needed a very different course of treatment if I were to arrest my illness. I felt so dreadful that I decided to take a few days off over Easter and I went with my family to Lamu. I thought that at the coast without anything to do I could rest and recuperate. We had a marvellous time and I began to feel as if I would again recover enough strength to go to Europe to plan the television films in detail.

While we were at Lamu I went out with the family to spend the day goggling and swimming. It happened to be a particularly calm day with very little wind so we took our small dinghy and went to a coral reef that was some way from the shore. The water was quite deep but clear enough for us to see a great deal. While goggling I decided to see how deep I could dive. I had never really been a very successful swimmer and I particularly disliked diving, but one of my friends had recently told me that you can blow out your ears as you go down, thus avoiding earache. I tried this technique but I found that the deeper I dived the more painful it was. I thought that perhaps I was not going quite deep enough, so I made a final determined effort and swam to a depth of, I suppose, twelve or fifteen feet. When I surfaced with a terrible pain in my head I found to my utter dismay that I could not hear a thing; I was virtually deaf. I supposed that the depth to which I had dived had resulted in water being forced into my ears and it would be only a question of time before my hearing came back. I swam over to the dinghy and explained my predicament to the family. They too thought that it was water in the ears and nobody took it very seriously.

After two days, we finished our holiday and returned to Nairobi. I was still almost totally deaf. I flew the plane myself and the only way that I could hear the aircraft radio when approaching the control tower at Nairobi was to use my earphones and to turn the volume of the radio to absolute maximum. In this way I could just carry out a conversation.

The following day I went to see my doctor, who peered into my ears and told me that the problem was simply one of wax – I had very waxy ears and the pressure of the water had forced plugs of it on to both eardrums. He gave me medication to soften the wax and arranged for the ears to be syringed out the next day. While I was there the doctor insisted, as he always did, that he take my blood pressure. He was appalled, it was the highest reading that he had ever had from me and it was beyond an acceptable level. He warned me that I was in a very serious condition and recommended that I should immediately see a specialist. I told him that I did not have time and, in any case, I wanted to delay the inevitable as long as possible. I asked him to give me some more medication to reduce the hypertension. He pointed out that I was already taking such a high dose of tablets that it would be very difficult to stuff any more in my system. We agreed, therefore, to change the type of medicine to see if such a change would have the desired effect. Meanwhile he had samples of my blood and urine tested.

After a couple of days I was feeling considerably worse and put it down to the new tablets. I returned to the doctor, who by then had the results of the tests and was in no doubt that I was in serious trouble. He advised me to go immediately to the hospital, which I did after completing my morning appointments. On arrival at the hospital I felt very depressed but I was soon given an examination to see if anything had gone wrong with my heart or other vital systems. My blood pressure on this occasion was so high that the nurse went off to fetch a second blood-pressure machine to verify the results. On getting a second reading that was above the scale she became rather alarmed and called a doctor – she had never heard of anybody living with such high blood pressure! Fortunately, I was in the care of an excellent specialist who soon got things under control.

When I was first admitted to the hospital in Nairobi, people inevitably heard about it. We tried to keep it a secret, but those in my office in particular were aware that I was very ill. Several people thought that I might have had a heart attack, but the suggestion taken most seriously was that I was the victim of witchcraft. At the time I was having a great deal of difficulty with certain members of my staff and there was an unfortunate undercurrent of hatred and bitterness in the Museum. There were a number of senior staff members who really did not like me and some of the junior staff thought that my hospitalization was due to a spell successfully cast by the former. Indeed, one of my more senior colleagues, a person who had received a PhD in America, was actually boasting openly that my illness was due to his particularly strong juju!

These stories caused consternation among those loyal to me because if I succumbed their turn would be next. In an attempt to alleviate the situation I allowed my friends to come to the hospital and treat me with an alternative witchcraft. They were sure that the only way I could possibly survive would be to receive appropriate traditional medicine rather than the western treatment that was being provided in the Nairobi Hospital. I duly saw a witchdoctor and various herbal medicines were brought to me. I confess that I was so ill that I did not really care what medicine I took although, needless to say, I did not tell my own doctor of these goings on. Here was I, supposed to be an intelligent person, receiving the most up-to-date modern medicine available in Kenya, while at the same time surreptitiously receiving traditional medicine from a powerful local Kenyan witchdoctor!

In addition to the things that were brought to me in the hospital, the witchdoctor insisted that he go to the Museum after working hours to put protective spells and charms in different parts of my office to keep out the evil spirits that now pervaded the whole building. I was in no condition to go around the Museum late at night, but it was acceptable for Meave to deputize for me. Meave duly met the witchdoctor one evening and spent a curious half-hour watching him put curses and spells in various parts of my room. I am not sure who believed what at this stage; we were all so desperately worried that we were prepared to do anything. It did however have the effect of improving the morale of my friends at the Museum which in itself was a good thing.

As well as trying traditional witchcraft, other friends subsequently suggested that we should try alternative sources of treatment. Some colleagues wanted me to treat myself with propolis, being convinced that this would be the solution to my deteriorating condition. Propolis, a compound made by honey bees, is undoubtedly very important as a treatment for certain illnesses, but it is a little late when the last stages of kidney failure have been reached. I was also invited to be a patient of a visiting German who practised homeopathy. I was to be tested with countless different sorts of natural extracts, my reactions to which would dictate the treatment.

I was persuaded to try one form of treatment which certainly relaxed me and made me feel better in myself, although I do not think that it improved my kidney condition. Twice or sometimes three times a week, a lady who is now a very dear friend of ours and who practises reflexology, spent half an hour kneading and rubbing my feet and toes. I do not know why reflexology is effective but it is certainly extremely pleasant.

My admission to hospital and the fact that it was now widely known that I was sick was a terrible blow. I was very upset and felt distressed and listless. Nevertheless within three days my blood pressure was reduced to an acceptably high level and I was released from hospital. But I was told that my blood pressure could go back up at any time and there was now an increasing need for me to be very careful how I lived my life. I began a somewhat half-hearted attempt at dieting, restricting my intake of protein, salt and various other tasty things that I enjoyed so much. After my discharge I did find that I was feeling better and so made up my mind to proceed to Europe for the meeting with the BBC team. I hoped that with the new drugs I might be able to keep going for at least another year or two.

On 16 June I set off for France and met Graham Massey and his team in Paris. We had a wonderful week looking at various prehistoric sites and eating the most delicious French food – none of which was on my diet! Sadly, my strength was visibly deteriorating. This became very obvious one day near the start of the trip when we were looking for a cave site. We were trying to locate a suitable place where we would eventually film a particular sequence. The rock shelter that we sought was down a small incline which could not have been more than 150 yards from the car and the total descent was probably less than a couple of hundred feet. It was the sort of scramble that a healthy person would do in less than five minutes. Going down did not seem particularly difficult and I scrambled after Graham and Peter enthusiastically talking about the view and the prospects for filming. Unfortunately, we found we had chosen the wrong footpath and the only way we could get where we wanted was to retrace our steps, cross over the valley and go down on the far side. We turned to go back up the hill and to my utter dismay, I found I simply could not keep up with the others. Because of my hypertension and the pains in my legs from fluid retention, I was unable to walk more than a few yards at a time without a rest.

I excused myself and proposed that, as I was not feeling too happy, the others in the party should continue and I would join them later at the car. This was my first admission to the BBC that I was in trouble. From then on I remained in the hotel or else in the car and we did our reconnaissance with four wheels. Inevitably, we had a discussion about my health. I explained that I had a kidney complaint, but that I was now being treated with new drugs and I was quite optimistic that although I would eventually have to have surgery or dialysis, this would not be in the immediate future. Graham seemed very doubtful that I would be fit for filming later in the year, but he considerately decided not to question me. He did ask if it would be appropriate for him to talk to my doctor but I said that I thought it would be better if my personal health were kept a private matter. Graham was clearly worried but we decided to continue with the plans as if nothing had happened.

I returned to Nairobi and on arrival I again, not surprisingly, had to be admitted to Nairobi Hospital. My blood pressure was so high that that in itself was causing further kidney damage and my kidneys were in such bad condition that there was no way that the blood pressure could be kept down. Until this time, none of the members of my family were really sure what was wrong with me but on this occasion they were told. The doctor let it be known that I was suffering from end-stage kidney failure. She explained that I would not be able to live on my own kidneys for much longer and that what I really needed was a transplant. Although this was the ideal way to proceed, if a transplant were not possible I could live on dialysis – an artificial system of cleansing the blood. There are no facilities in Kenya for kidney transplants or for lengthy dialysis programmes, so I would have to go abroad. In England it would not be possible for me, as a Kenyan citizen, to receive a kidney graft from a dead person; in the United Kingdom kidneys from cadavers are quite rightly kept for British patients. In view of this I had the option to go to another country where cadaveric kidneys are available to any patient.

I realized that there was a good chance that I would have to live on dialysis for the rest of my life. The prospect of this was far from encouraging, although people do live on artificial kidneys for many years without any difficulty. The principal problem is that one is not able to be very active because of the need to be at the machine every second or third day. It is simply not possible to transport a large kidney machine around a country like Kenya where water, electricity and sterile conditions are unreliable. Knowing all this was depressing but I was tremendously flattered to be approached independently by my three brothers who each offered one of their kidneys. To receive a kidney from a living donor is by far the most satisfactory procedure for transplantation, but long before a doctor will take an organ from a living donor, the medical team has to be completely satisfied that the chances of rejection are minimal: the tissue match between the donor and recipient has to be essentially perfect. My two brothers in Kenya, Jonathan and Philip, agreed to send blood samples to London for tissue typing while my half-brother, Colin, who lives in England stood by in case he might be needed. Colin was my father's son by his first marriage and he is eleven years older than me. He and his sister

Priscilla had been brought up in England and it was not until 1963 that we really got to know them as relatives. In addition to the offers from my brothers, I also received offers of a kidney from a number of friends whose tissues would almost certainly be unsuitable. The fact that they were prepared to give up a part of their body to keep me alive is quite extraordinary and I remain forever grateful. It is impossible to describe the generosity, concern and love that I received from so many friends at this difficult moment of my life.

Following my release from hospital, I had to make daily visits to the doctor for checks. I was swallowing twenty-six pills daily and I was permanently drugged. The uraemia that I was suffering from made life very unpleasant and I knew that I was increasingly difficult to cope with. Meave and several close friends understood what was happening and gave me tremendous support but I was still unable to believe that I had reached the end of the road; I was telling myself that given a few more weeks on the new drugs I would recover sufficiently to film with the BBC in July. I was so sure that I could do this that I accepted a suggestion from my doctor that in early July I should go to England to see a specialist about the hypertension. I was quite satisfied that the purpose of the visit was to get my blood pressure under control and I believed that with reduced blood pressure, I would be able to string out my life a little longer on my own kidneys. How naive I was! I proposed to go alone; after all, I told myself, I was only going to pick up some extra tablets that would keep my blood pressure down; and it would be a short visit; I would be back in Nairobi in less than ten days. The drug that the doctor wanted me to have was not available in Kenya – it was still experimental – so I just had to go to England.

Meave insisted on coming with me and I suppose I should have realized then that she knew just how ill I actually was. I argued with her that it was an unnecessary expense and that I would only be gone a very short time but she was resolute and secretly I was thankful. I knew in my heart that I could not manage without her.

CHAPTER FOURTEEN

THE END OF ONE LIFE

WE MADE ARRANGEMENTS TO DEPART for London late at night on 13 July following a small party at our house for a number of friends. Each of our guests knew us well and although there was little to suggest that it was a gathering to say farewell, I was keenly aware that this was so. I suppose everyone had got used to my condition and nobody drew attention to it. There are a number of unpleasant symptoms of kidney failure which are difficult to describe adequately. I felt desperately cold from within; I was quite unable to get warm and even when I was wrapped up in woollen sweaters and seated by an open fire I still felt 'steel' cold. Another symptom of uraemia is nausea. I constantly felt I was about to vomit; indeed, I was vomiting four or five times a day. I had a great tendency to bleed from the nose and any nose blowing resulted in a steady trickle for some minutes. The taste associated with uraemia must be the most unpleasant feature; I am quite unable to define the sickly but bitter taste that was present in the back of my mouth and in my nose. It lasted for months and varied in intensity as did a constantly itching skin. My back and chest were a mass of irritation which made relaxed sleep difficult.

Because I insisted that it was a short trip, we took very little with us – just two overnight bags. We left our children with relatives, and friends were to look after the house. When it was time for us to leave home, we bade everyone a rapid farewell, begged our friends to continue the party, and got ourselves into my car to be driven to the airport by my driver. As I sat beside him in the dark, I was startled to find that tears were quietly trickling down my face and any attempt to say a word would have given away the fact that I was weeping. I do not know exactly what Meave was thinking but I realized that I might well never return to my beloved home or see again my children or friends. I tried hard to maintain a false cheerfulness and confidence because I knew that if I gave way to my innermost feelings I would never keep going.

The flight to London was uneventful; I slept all the way as I normally do. On arrival, we were met by a friend, Hazel Wood, who took us to her home for the day before moving us to another friend's flat which was to be our base for the week or so that I planned to be in London. I still hoped, and perhaps believed, that I would get better!

We had arrived in London on a Sunday and the next day I went to a laboratory to leave a blood sample for tissue typing. This, of course, was the first essential step. My two Kenya-based brothers sent blood for analysis on the Sunday night-flight from Nairobi, and Meave went to London airport to meet the Kenya Airways plane whose pilot was carrying the containers. After leaving the laboratory I went immediately to the consulting rooms close by St Thomas' Hospital where I met my physician. He examined me thoroughly and blood tests were made. I was asked to return the following day for further checks and a full discussion on the results of the initial tests. When I returned to our London flat I had a strange feeling that the doctor had rather ignored my conversation about going home and I made a mental note to be a little firmer on the next visit. The rest of that day I spent sleeping. I was feeling much worse than even a week before and I was passing blood whenever I vomited. My gums were also bleeding and some small blood vessels in my eyes had ruptured causing bright red patches in both eyes.

On my second visit to the doctor, he was quick off the mark and before I had a chance to discuss my intentions I was told that he wanted me in hospital at once for observation. I was told in Meave's presence that I had end-stage renal failure, that I could only live if given artificial means or a transplant and that no further delay was conceivable. It was an incredible blow, even if secretly expected. I wanted to see my children and I so wanted to see Kenya just once more! Meave, however, pointed out that I really was far too sick to travel and that the flight alone would do me more harm than good.

One of the first things the doctor arranged was for me to be shown the renal ward at the hospital. I met the sister in charge of the unit and several patients, some of whom had been on dialysis for a good number of years – all of which I found very reassuring. Some patients looked remarkably fit and I was told that they were able to live reasonably normal lives; they had regular jobs and often came in for dialysis at night on a schedule of three times a week. This opportunity to see living proof that I could survive even if a transplant was unsuccessful was tremendously encouraging.

My doctor explained to me that even if I were able to arrange to have a kidney transplant immediately, I would need dialysis to keep me alive. Several weeks of this would be necessary to make me stronger and better able to cope with the operation and any subsequent complications. Before dialysis would be possible, I would require a small operation to prepare a suitable point of access to the blood supply. This would take place while I was in hospital under observation. The operation involved minor surgery to connect an artery and a vein in the forearm, thus forming a 'fistula'. The veins in the forearm swell with the increased flow of blood and needles can be inserted into these veins to draw off blood and return it during a dialysis. Haemodialysis involves passing blood through filters or membranes which remove the various impurities normally dealt with by the kidneys. These filters are 'artificial kidneys' and it is necessary to circulate the blood through them for between five and eight hours three times a week. Unless the blood can be drawn off at a good flow rate, the process takes even longer. Most of the body's easily accessible veins simply do not have a large enough blood flow, hence the need for a fistula.

As a result of the long illness my general condition was poor, so before I could have any operation I was given some blood by transfusion. This had a tremendous effect; I felt so much stronger and alert that I began to realize how good it was to be well. Over the three or four hours of the transfusion, I could actually feel myself changing for the better. During the following four months I was to have many transfusions and I always looked forward to them because of their immediate effect. Previously, I had never really appreciated how valuable blood donation can be but having been so dependent upon other people's blood for so long, I salute the many unknown donors.

Perhaps the most unpleasant shock that I had when I went into St Thomas' for the first time was the impersonalization; until then I believed that I had complete control of my affairs, but in this new and strange English hospital I felt very much alone. I had always been aware that doctors and nurses have to treat the human body objectively but to have personal experience of this necessary attitude was quite new. The poking and prodding, the investigations and questions soon reduced me to being just a patient and I was very miserable. It seems to me that the process begins as soon as you have to remove your own clothes to put on hospital garb. I always felt that it was Richard Leakey being put on the hangers in the cupboard and sometimes I would talk to him hanging there while I lay in my bed! Worse of all was to see these clothes being taken away making it clear that I was not going to be needing them in the near future.

The fistula operation was done by the surgeon who was later to perform my transplant. He was greatly feared by nurses and junior doctors alike and I was quite interested to meet the man who was going to be so critical to my future. As it happened we got on very well and I was to benefit tremendously from his direct and authoritative manner. The fistula operation was completed and in a few days I was released from hospital.

One of the first things I had to attend to was my obligation to the BBC.

The producers, Graham Massey and Peter Spry-Leverton, were wonderful. They did not show even the slightest annoyance as they might well have done, considering the predicament in which I had placed them; a huge budget had been approved, a professional crew was assembled and a few months before starting to film, their central figure drops out – perhaps permanently! I made every effort to reassure them that it would not take long for me to get fixed up and we agreed that filming should begin without me but that I would join the team as soon as I could. It amazes me that they went on with the project. Their confidence in me was extremely helpful because the prospect of my joining them and completing the series helped me through many of the difficult times that were to follow.

By the time of my discharge from hospital, Meave and I had been away from home for more than a week and we both realized that we would not soon be returning to Kenya. Meave decided that the children should join us and even though I wanted this myself, their coming depressed me further for a while because it confirmed my changed situation. I was not likely to be going back to my former life for some time, if ever, and each reminder of this grieved me. The good news was that the tissue typing had been done, and my younger brother Philip had what the medical profession calls a 'full-house match'. His tissue type was identical to mine. The doctors guessed – correctly – that he would bear the closest physical resemblance to me though there are, in fact, four years between us. Identical tissue types are a feature of identical twins and there is a one-in-four chance of a brother showing identical tissue with his sibling.

Curiously enough it was by no means a matter of course that I decided to accept such a gift from my brother. Knowing that Philip was a suitable match, and consequently the prospects for organ grafting ideal, relieved me but this was only part of the problem. The big question was, should I accept his offer? I had to come to terms with the idea of receiving an organ from a brother or, indeed, from any living donor. The fact that my brothers and several friends were prepared to put their own lives in danger by offering me a kidney was overwhelming. Voluntarily to submit to major surgery and to give up nature's 'failsafe' system of two kidneys is an act of sacrifice which is almost impossible to evaluate. My problem was to accept the responsibility of allowing such a sacrifice.

I had many misgivings. What would I feel should my donor fail to recover from the operation itself? Could I live with my own conscience if my body were to reject the grafted organ, so making the whole gift a tragic waste? Such rejections are known and there was a real chance, which could not be overlooked, that this might happen. What if Philip were later to be involved in an accident where damage to one kidney would prove fatal in the absence of a second? Yet I desperately wanted to live again a full life – to be an active father to my family, to walk in the deserts, and to swim in the oceans. A transplant was the only possibility. As I have said, I had no opportunity of having what is known as a cadaveric graft in England and I felt strongly that the British medical authorities' decision to restrict donated organs to British subjects was absolutely correct. I could have considered treatment in another country where private medicine does not impose these restrictions, but having gone to London, it was difficult to change at the eleventh hour.

My dilemma over whether to accept a kidney from my brother was made much worse because for years we had not been close. In fact, we had been as distant as two brothers can be. We had seriously disagreed several times about ten years before and since that time, we had had absolutely no contact. Indeed, it could be said that I was distinctly unfriendly to him and his family; for more than a decade none of them had been invited into my house. Although the reasons for our bad relationship are unimportant here, it must be said that I was very conscious of this and it seemed wrong to use him just because I wanted to live. Equally, I had my pride and the idea of being befriended by Philip simply because I was ill was hard to accept. It was an extraordinary act of generosity that, on hearing of my illness, Philip immediately offered me one of his kidneys and I was not sure that I was big enough to accept this gesture graciously in the spirit in which it was made.

My being in England with Philip in Kenya may have made things more difficult. I had, of course, spoken briefly to both my brothers when I was in hospital in Nairobi just before I was sent to England. The discussion had been uneasy and I fear I was quite ungracious when offered help. At the time I think I merely said that I hoped it would not be necessary to have to take up their offer, but thanks anyway! As it turned out, I was now facing the biggest decision of my short life and I had not the benefit of good health which would have enabled me to think rationally.

Meave and I discussed it many times and my desire to get better was always there, but the decision had to be my own and I could not expect help from outside. Our good friend Bernard Wood arranged for me to talk to another specialist; I wanted a second opinion on the prospects of a transplant. I suppose that I was subconsciously seeking reassurance and hoping to find something that would make my decision easier. With this specialist I reviewed the prospects for life as a permanent dialysis patient and I was given all sorts of statistics. The prospects for a successful

ABOVE An aerial view, at sunset, of the camp on the sandspit at Koobi Fora

LEFT Dry river beds are generally the best place to look for water and, even in desert areas, there is a surprising amount of it if you know where to dig

ABOVE Trying my hand at re-assembling fossil fragments with advice from Meave, who is holding Louise

LEFT The Masai of Olduvai are colourful people and often inclined, in an impromptu dance, to see how high each can jump. Even their small children join in transplant were gone over again as were the known statistics for donor survival. I was given a little book that had been written by a Sudanese patient who had had exactly my problem. All of this helped and yet I was still very reluctant to accept the gift of Philip's kidney.

I finally wrote Philip a letter setting out my worries and asking that he seriously reconsider the whole offer in the cold light of day. I explained that Meave and I could accept a change of mind in advance of the event but that we could not accept the chance that there might be regrets after the transplant. At the time, Philip was busy preparing his election campaign to run for Kenya's Parliament and I am sure that my complex questions were not especially helpful. Philip replied at once to my letter stating simply that he had meant his offer to be taken as a serious one and that there was no need for any reconsideration. He would be in London as soon as the elections were over. His letter was short and to the point and it made my decision much easier. I accepted and wrote to thank him without further ado. From this point on life took on a new meaning.

I remember very little about the weeks in England before dialysis began. I spent most of each day sleeping and the whole period became a long nightmare. At this stage, my kidneys simply weren't coping and I was retaining a lot of fluid. I was urinating, but not enough. The fluid retention became so bad that I was forced to sleep propped up because otherwise my lungs would become awash and I would begin to choke and drown. I am told that there were many friends who visited us but although I was dressed and lying on a settee, I was often quite unable to awaken or if awake to concentrate on the people who were present. I would start a sentence in a conversation and never finish. I found any physical effort difficult and a walk of a mere 50 yards to sit on the park bench near where we were living was a major undertaking requiring several long rests on the way. My condition was a terrible burden to Meave, who had to watch helplessly as my mental and physical state deteriorated. I have always made light of all but the most trivial ailments and I tried desperately to convince everyone that I was not feeling as sick as I actually was.

I am not entirely sure why I was unable to start dialysis any sooner than I did. My fistula was one problem because the operation had to be redone, and perhaps the shortage of kidney machines was another factor; I was waiting my turn for the very limited number of places in the renal programme. I also think that there may have been an important psychological reason; dialysis is not particularly pleasant and many patients become depressed by the process of being tied to a machine. The sicker a patient feels at the beginning of the treatment, the more dramatic the improvement. In my own experience it was perhaps a good plan to have gone so far down because I was truly delighted by the effect of dialysis when it began.

I went into surgery for the second fistula operation following another wonderful gift of blood. There was no doubt that this fistula was working; it was very dramatic because it made a tremendous noise for several days. This may sound very odd but the blood being pumped back from an artery directly into the large vein in my forearm made a loud pumping noise which was clearly audible from some distance. As it happened, I went with Meave to have dinner at the Dorchester as the guests of some important American friends the day that I was discharged from hospital. As we sat at a table in the famous hotel, my arm distracted our host and hostess as well as several passing waiters!

At the end of August, a few days after the operation, I began dialysis. I was pleased to begin what I saw as the start of the long road back to better health. I knew not the outcome but I was quite sure that the worst was over and I looked forward to the unknown. When I arrived at the renal unit I was in for a surprise because dialysis was to be done from access to blood vessels deep in my groin rather than from the now loudly pumping wrist fistula which was not yet properly healed.

The 'groin dialysis', as I call it, is quite a different matter from the standard procedure. A doctor has to find the femoral vein which is deeply buried in the groin. After a local anaesthetic, a long thick needle is pushed into the groin until the large vein is found and entered. A steel wire is then pushed through the needle into and along the vein for five or six inches. A plastic tube is then run down the wire which is subsequently removed and the tube is attached to the kidney machine. The same process is repeated on the other groin because one tube takes the blood to the artificial kidney while the other returns the blood to the body. Naturally, this is all fairly uncomfortable and I lay on my bed wondering what would happen next. I spent ten hours on the machine that first time – as far as I was concerned it could have been for longer because for the first time in months my uraemia was gone. That terrible taste in my mouth was practically forgotten and there was the beginning of a new warmth in my body.

Because I had expected to be feeling so much better after my first session on the machine I had invited a friend, Nonnie Kellogg, to come for dinner that evening. I particularly looked forward to seeing her and hearing first-hand news from Kenya – she was returning from Nairobi to America after a month's holiday. As it turned out it was a good plan because Meave was able to leave the children with Nonnie while she came to collect me. I had hoped to be able to get myself to the flat in a taxi but it proved more difficult than I expected.

The first dialysis reduced the excess fluid in my body and my weight loss was two and a half kilograms or about five pounds. This left me feeling very weak and I was tremendously surprised to be unable to get out of bed to dress or to leave under my own steam. Meave had to take me through the hospital in a wheelchair to a waiting taxi and help me into our flat at the end of the journey. I was, nevertheless, very pleased with myself because I felt so much better. It seemed especially right to be able to celebrate with Nonnie who had patiently waited for several hours for her dinner. I returned for dialysis two days later and the whole procedure with my groins was repeated. At the end of the third session, my groins, being bruised and very tender, were a brilliant colour. The idea of a fourth session was out of the question and it was decided to begin using my fistula. Fortunately this worked very well and the painful aspect of my dialysis treatment was over for ever. The sutures on my fistula scar were actually removed after dialysis had started.

By the end of the second week of September I was well established with my hospital routine and I began to see ahead with greater confidence. I saw the dialysis treatment as a holding action in preparation for the transplant which I hoped would provide me with a second life. It was difficult to plan the transplant because Philip was still waiting for the elections in Kenya and the polling date had not been decided. The plan was that he would come to London immediately after the results were announced. All this time I began to feel increasingly concerned about the financial aspects of my illness because I had only limited resources. Treatment for a private patient in England is not inexpensive and in addition we were spending quite a lot on simply living in London. Fortunately, the National Museum had undertaken to make a substantial contribution towards my medical costs, but even with this I was worried; if Philip's arrival were delayed we could have another crisis. We had hoped, initially, for an August operation and on this basis I had begun to make a filming schedule with the BBC. I was not taken too seriously and looking back on it all, there must have been every sign that I was going slightly crazy. However, my friends were all very kind and nobody ever suggested that I would not be up and about by the end of September. Days came and went and each day I eagerly hoped for news of Kenya's elections.

While I was having problems in London, the Museum in Nairobi was undergoing a difficult time too. Before I left home, it had become very clear that there was a very serious effort being made to remove me from my job as Director of Kenya's National Museum with responsibility for Antiquities and Monuments. The moves to displace me had really begun in 1978 and had reached their height by March 1979. Knowing this before I left, I had been able to take certain measures to contain the situation and these had involved a range of discussions and agreements with senior government officials. My great concern had been to ensure that those who were against me should not take undue advantage of my deteriorating health and absence from Kenya. The whole problem related to the growth of the Memorial Institute which by early 1979 was a fairly independent part of the National Museum. Those who were against me wanted to separate the various functions of the National Museum and by so doing, gain control of the Institute and neutralize its effectiveness.

I was especially bitter about this because from October 1968, when my career began with the Museum, the organization's growth had been dramatic; the number of staff rose from 23 in 1969 to around 300 (including some casual employees) in 1979, while the budget had increased over the same period from a mere £23,000 to over £600,000. All of this had been achieved with considerable effort and my great ambition had been to complete the Museum expansion before I succumbed to what I knew was inevitable: kidney failure. At the very moment when success was in sight, I was facing a distinct possibility that the organization I had built was to be broken up and I was not in a position to fight back.

Among the various proposals, it was agreed that the Kenya Government would set up an independent commission to examine the National Museum and make recommendations to the Government about how things should be run in the future. My hopes lay in the integrity and common sense of the commissioners; I believed they would vindicate my opposition to the moves made by my opponents during the last few months of illness. All my staff knew that should I fail to get this vote of confidence, I must lose my position as Director. By the same token if, as I hoped, the Government accepted my policies as a basis for the future of Kenya's museum system, then I could reasonably expect my detractors to move on to other things themselves. All of this was being actively discussed while I was on dialysis and I spent a small fortune on international telephone calls trying to keep abreast of developments. As it happened, I was worrying unduly because I enjoyed the full support of the Minister in the Government who was in charge of museums. He, in turn, had received a very positive report from the commission. It was soon clear that if I could regain my health, my job was safe.

While I was in England undergoing treatment, news of my illness spread and there began a flood of letters. The sheer volume of this correspondence was very flattering and it gave me plenty to do as I attempted to answer every letter. I was particularly grateful for the many encouraging letters I received from our close friends; this correspondence was tremendously important to me. Some of the other mail came from people who only knew of me through my work and it really was quite extraordinary to receive letters from virtually every part of the world. Many of these well-wishers clearly did not know the details of my illness; they referred to me almost in the past tense which led the family jokingly to call the correspondence 'the obituary mail'. On one occasion I even received a press cutting which spoke of the 'late Richard Leakey'! All of this provided a welcome distraction from an otherwise rather difficult existence.

We had acquired a flat which we rented for the duration of our stay in London and both Louise and Samira attended a local school which was about ten minutes' walk from where we lived. I had to go to the hospital on Mondays, Wednesdays and Fridays, usually arriving at the renal unit at about eight o'clock when the morning shift of nurses began. By choice I liked to start early because this enabled me to get back to the flat for a useful afternoon. For no good reason, I found a regular schedule far less demanding; I simply hated to allow my days to be unstructured even though I was doing nothing of consequence.

The dialysis routine itself was quite straightforward. Upon arrival at the unit, I was allocated a machine and if possible this was located in a small side room where I could be alone. I particularly appreciated the privacy because, in addition to my writing, I used to see quite a number of visitors. My first task on arrival was to prepare the machine. This involved attaching the various parts of the 'disposable' equipment: the artificial kidney itself, the blood lines, needles, clamps, a saline drip bag and various other items. It usually took about fifteen minutes to get everything ready although some days it seemed to take longer because I could never work out the flow circuit. However, once ready, I had only to weigh myself before asking one of the nurses to attach me. This was the process of inserting the needles into my vein and releasing the clamps so that the blood could begin to be pumped through the artificial kidney. I could have done this myself but it seemed at the time that it would have been one step too far in accepting my situation.

Once the machine was running, I was able to settle down to work. Rather than be dialysed in a bed, I opted to sit in an arm chair so that I could write at a table using my free right arm. It was a successful arrangement and I was able to complete a fair volume of hand-written correspondence as well as most of the manuscript for this book. The idea of simply lying in bed and reading was most unattractive.

One of the least welcome distractions was to find that I was becoming too dry from dialysis and so 'flopping' as it is called. The process of dialysis removes fluid along with impurities in the blood, and if too much fluid is taken off one begins to feel faint and distinctly nauseous. The only remedy is to put some fluid back into the system and the easiest and quickest way is by means of a drip. Flopping is a common occurrence and I was shown how to be prepared. I used to set up a bag of saline solution before beginning my dialysis. As soon as I felt my powers of concentration going, I would simply reach over and release a clamp, thus allowing fluid to flow straight into my blood system. After several weeks, I was able to devote my entire dialysis period to writing without any major loss of time.

I found the most distracting aspect of dialysis was the noise of the machines. Each of them is designed with an extensive system of alarms which registers any fault. For example, if the blood flow is interrupted because a blood line gets blocked or if air enters the system, a loud bleating alarm soon attracts a nurse. With fifteen machines operating in the unit there was an alarm sounding somewhere every ten or fifteen minutes. I thought that the sound of these alarms was the most distressing aspect of the treatment. The noise is a desperate reminder that each patient is tied to an artificial lifeline and I wonder if the makers of these machines might not think of some other means of alerting the nursing staff, such as a signal on a remote control panel. A musical tune would be better than the shrill bleating which sounds so like a siren with its portent of disaster.

At the end of a six-hour session, a nurse would take me off my machine. This involved clamping off the lines after all the blood in the machine had been returned to my body. Once the needles and lines were removed it took only a few moments to stop any bleeding and after a weigh-in at the scales for the records, I was off home. I always felt so much better after dialysis. I suppose the hardest part of dialysis is the strict regime of liquid intake. Because the body cannot rid itself of fluid except through the kidneys and sweating, kidney failure means that almost all liquid taken in has to be removed by dialysis. With this situation, a patient has to be extremely careful to avoid drinking too much; excess fluid is both uncomfortable and dangerous. I was put on a ration of about 0.6 of a pint per twenty-four hours, which is equivalent to just over two tea cups. Sometimes I would have to cheat and have three cups but this never even got near to quenching my thirst. To feel thirsty all the time for three months was a great discomfort for me and I recall the extraordinary pleasure provided by each mouthful of liquid I took.

As a dialysis patient I was told to be very careful of what I ate, and to keep well clear of certain foods such as fruits, mushrooms and a host of other delights. Some of these foods have a high concentration of potassium that can build up to dangerous levels in the blood. I did my best and fortunately it was good enough, although I sometimes stretched the rules; my view was that a little pleasure was as important to my survival as anything else.

By the beginning of October there was still no definite news of Philip's date of arrival and I began to get depressed. I seriously considered the option of remaining a dialysis patient and again examined the problems of home dialysis in Kenya. At the time my great desire was to get away from England; to return my family to their own land and to be with my own people again. It is not that we were uncomfortable, our existence in London was fine – I just wanted to enjoy the familiar sounds and smells of Kenya again. Meave consoled me with the thought that a three-month wait was nothing compared to what many kidney patients have to suffer. Although I was able to appreciate this argument I was still very impatient and fretful. Perhaps it was because the transplant was such a critical event that I felt a desperate need to know whether it was going to work or not. Each day dragged to the next and there was still no news.

In mid-October the date of the Kenya elections was finally announced, and a firm schedule began to take shape. Polling day was to be 8 November and I could look forward to my operation shortly after that. The prospects for geting home suddenly seemed much brighter and I began to smile again. It is not possible to guarantee the success of a transplant before the operation, nevertheless I had absolute confidence that whatever was to come, at least it would be definitive and this was important. I knew that if the transplant failed, I would then have to commit myself to a different life once and for all. I felt certain that I could successfully do this which meant that I was somewhat fatalistic about the impending surgery.

Meave and I 'phoned Philip a number of times to discuss the final arrangements for his coming over to London. I wanted to be sure that as little time as possible was lost but there were other considerations. It was essential to give the doctors and transplant team a firm date so that the operation could be scheduled to suit their calendar of other work. Setting a precise date proved to be extremely difficult because there were two quite different prospects; if Philip lost his bid for a Parliamentary seat he could come immediately after the poll, but if he were elected, he would have to delay his departure for several weeks so as to be able to complete all the post-election formalities. We settled for a date in late November which allowed for the eventuality of his winning. Philip agreed that he and his wife Valerie would fly to London on 25 November to be ready for the operation on 29 November.

The final few weeks before the operation were perhaps the most difficult for me to deal with. I became quite preoccupied with the possibility of something going wrong in the final stages. I had visions of Philip being unable to come over because of an accident, or worse, perhaps he would find himself the target of an assassination bid in Kenya where electioneering can be fairly rough! With all of these fears on my mind I was very much on edge.

Philip was competing against the incumbent M.P. as well as nine other candidates. His constituency, known as Langata, is a very densely populated suburb of Nairobi, and his constituents were primarily citizens of African origin. His candidacy was a point of great interest because, Philip apart, there had never been an attempt by a 'European' Kenyan to enter politics since Kenya had attained its independence from Britain in 1963. Philip had run for the Langata seat in 1974 but had lost in a closely fought race. In the 1979 election he won with a comfortable majority.

Everyone was delighted by his success although I confess I had a most selfish moment when I reflected upon the delay to my operation that this victory had caused. For Philip to leave Kenya shortly after winning his seat was an act of great sacrifice and many people were absolutely amazed that he was willing to go off to England for surgery at this point in his career. Nevertheless he did and he and Valerie arrived in London as planned on 25 November. They phoned up shortly after arrival and we agreed to meet for lunch that day.

This meeting was the very first occasion that the four of us had ever sat down to a meal together. I think we were all conscious of this and looking back on it, it is hardly surprising that we were somewhat stiff. During lunch Philip told us all the news from Kenya and in particular the details of his election campaign. It was exciting to feel that soon we should all be back home and our conversation eventually turned to the arrangements for the operation. At that point I had a major surprise: Philip calmly informed me that he could not go into hospital until such time as he was able to arrange for police protection! He had information that an attempt might be made to kill him while he was in the hospital. It was all a little vague but he and Valerie were quite adamant. Philip also pointed out to me that since we look so alike I might get assassinated by mistake!

To arrange for police protection in one's own country is difficult enough but to persuade the British authorities in London that they should help out in such a situation is very much worse. There was nothing for it but to try and so I set to work on the telephone to make the necessary arrangements. Fortunately I have a number of good friends in London and in due course the arrangements were made. There was an inevitable delay and I well recall the reaction of my doctors when I phoned to say that Philip would be held up for twenty-four hours because we had to arrange for armed bodyguards at the hospital!

Philip finally went into St Thomas' hospital on 27 November and underwent the tests that are always carried out just before the removal of a living organ for transplantation. I went there at noon the next day and we were put into adjacent rooms in the private wing. Philip joined me for lunch and we sat together on my bed having the second meal that we had had together as adults. The next day he was to give me one of his kidneys. I was still somewhat concerned and I again raised the question of whether or not he was absolutely certain that he wanted to go through with the arrangement. I was terrified that he might back out but I felt I had to be quite sure.

My mother had also come to London and she was as anxious as any mother would be. She was also, I think, a little nervous about the relationship between her two sons because for so many years she had had to deal with us individually. The sight of us together and apparently getting on well with each other was novel. One friend had in fact said that the relationship between Philip and me had been so bad for so long that my body would almost certainly reject Philip's kidney after the transplant.

On the day of the operation we were all a little on edge. From my point of view, the sooner the whole thing was over the better. I wanted to know whether or not I was to begin a new life. Towards noon the preparations for surgery were begun and by 1.15 p.m. we were both on our trolleys and on our way down the corridors of the hospital. I tried to talk to Philip while we were wheeled along but it was difficult as both of us had been given sedatives. In the anaesthesia room, immediately outside the operating theatre, I requested that my trolley be taken alongside Philip's and I asked him again whether he wished to go through with things. His reply was short, blunt and unprintable but it reassured me and gave the nurses and others present a moment of laughter. I, too, laughed and that was the last I remember until late that night. I awoke to the sight I had so longed for: hanging on my bed was a bottle containing a rather bloody liquid which I immediately recognized as my urine. My second life had begun.

POSTSCRIPT

MY RECOVERY FROM THE TRANSPLANT surgery was rapid and in a few days I began to feel very fit. For the first time in months I could again eat and drink as I wished and as I refused hospital fare poor Meave had to spend hours preparing and bringing me meals from our flat. After only thirteen days I was released from hospital. Philip had left hospital three days previously, and a month after the operation he was back in Kenya. Since then he has been none the worse for his donation.

I had been warned that rejection of grafted organs is common and that I should not presume that I was fully recovered for a number of months. But as the days passed and my kidney function became virtually normal, I began to be more and more optimistic that I had one of those rare but possible 'perfect' transplants. How wrong I was. On Christmas Eve, three weeks after the operation, I began a massive rejection episode and by Boxing Day, my kidney function was back to zero. The shock was unbelievable. I was readmitted to hospital, where I was treated with immunosuppressive drugs to halt the rejection. The doctors were worried that the kidney had been permanently damaged by the rejection, but I was lucky. Gradually it began to function again and on 29 December I was discharged, but my problems were not yet over. The massive dose of immunosuppressives had left me with no immunity against infections. On New Year's Day I went back into hospital with a very high temperature which turned out to be a rather unpleasant viral infection. I was kept there for ten days and there were moments when I felt that my second life was going to be rather short! Once more I returned to the flat and again looked forward to an uneventful convalescence but the worst was yet to come. In the early hours of 20 January I was rushed back to hospital, this time in an ambulance; I had suddenly gone down with severe pneumonia complicated by pleurisy and septicaemia. I was very ill and Meave was warned that my chances of recovery were uncertain. Thanks to a magnificent effort by doctors and nurses I can write this postscript, but I should say that had it not been for Meave who gave me the strength and hope to fight, I would not have lived. We eventually returned home to Kenya on 23 February 1980 and I have enjoyed perfect health ever since.

PICTURE CREDITS

With the exception of the illustrations credited below, the photographs in this book were taken by the author, his wife, members of his family, and close friends. In a few cases the photographer is now unknown. To all, amateur and professional, the author and the producers extend their gratitude.

BLACK AND WHITE

p.13 The map was drawn by Eugene Fleury

spread between pp.40 & 41 (centre left and bottom left): British Airways

facing p.56 (bottom left): Charles Trotter

spread between pp.56 & 57: 'Daily Nation', Nairobi

facing p.64: Mary Griswold Smith

facing p.65 (both): Bob Campbell

facing p.80 (all): Bob Campbell

spread between pp.80 & 81 (top left): Winfield Parks © National Geographic Society; *(bottom left, all; top right):* Bob Campbell

facing p.81 (all): Bob Campbell

facing p.96 (top): © National Geographic Society; *(bottom left);* Marion Kaplan

facing p.97 (bottom, both): Marion Kaplan

facing p.112 (top): Ian Findlater

facing p.113 (top): Pippa Copp/BBC; *(bottom):* Marion Kaplan

facing p.128 (top): Marion Kaplan

COLOUR

facing p.88 (top): Bob Campbell

spread between pp.88 & 89 (both): Bob Campbell

facing p.89: Bob Campbell

facing p.104: Joanne Hess

spread between pp.104 & 105 (left): Glynn Isaac

facing p.105: Bob Campbell

facing p.136 (both): Bob Campbell

facing p.137 (both): Bob Campbell

facing p.144: Bob Campbell

facing p.145 (both): Bob Campbell

facing p.160 (both): Bob Campbell

facing p.161: Bob Campbell

spread between pp.168 & 169 (bottom right): Bob Campbell

facing p.169 (bottom): Kay Behrensmeyer

facing p.176 (top): Heather Campbell

facing p.177 (top): Ian Findlater; *(bottom):* Bob Campbell

facing p.192 (top): Bob Campbell

facing p.193 (top): Bob Campbell

INDEX